Farmers Market Manual

Kristin Brennan

Contributors:
Sara Coblyn
Colleen O'Brien

THE FOOD PROJECT
Lincoln and Roxbury, Massachusetts

United Book Prints

Reprinted in India by:

UNITED BOOK PRINTS
5A, Bhagat Ki Kothi,
Near Bhagat Ki Kothi Police Chowki
Jodhpur 342 001, India
E-mail: unitedbookprints@gmail.com

Print : 2021

Designer: J.A. James
Editors: Carlos Christensen and J.A. James
Photographs: Greig Cranna, John Walker, and The Food Project Staff

ISBN: 978-93-88399-00-5

Printed in India

Our Vision

Creating personal and social change
through sustainable agriculture.

Our Mission

The Food Project's mission is to create
a thoughtful and productive community
of youth and adults from diverse backgrounds
who work together to build a sustainable food system.
Our community produces healthy food
for residents of the city and suburbs,
provides youth leadership opportunities,
and inspires and supports others
to create change in their own communities.

Preface

The Food Project started in 1991 in Lincoln, MA, a wealthy Boston suburb, on two and a half acres of farmland. It was a small, noisy and energetic community of young people from very different races and backgrounds, working side by side with adults growing and distributing food to the hungry.

In the process of growing food together, we created a community which bridges the city and suburb, is respectful and productive, and models hope and purpose. We teach how one can love the land and its bounty and how to care for and respect the rich matrix of life to which we all belong. In doing this we as individuals and as a society grow and develop in healthy and sustainable ways.

We have grown since 1991 and now we farm on thirty-one acres and over two acres of remediated land in Roxbury, a low-income neighborhood in Boston. We provide year-round stipended programs for over one hundred youth, and with the additional help of 1,700 volunteers, grow 200,000 pounds of food for fifteen shelters, our two farmers' markets in low-income neighborhoods, and over two hundred Community Supported Agriculture shareholders. Since 1995 we have played an important role in increasing productive green space and access to fresh food in Roxbury through remediating idle land, and mentoring neighborhood gardeners in healthy food production. Food Project alumni play critical roles in shaping program growth in enterprise, community organizing and education for urban gardeners, and building a national network to engage youth in local food system creation. Alumni interns are agricultural apprentices, support our technology, pilot food enterprise with our chef, and play pivotal roles in outreach and education. Alumni also serve as Food Project staff, trustees, and advisors.

The Food Project's inspiring model addresses critical national issues: the need for race reconciliation, the decline in local agriculture, a growing concern for the well being and productivity of youth, and the need to create sustainable and healthy inner-city neighborhoods and metropolitan areas. The Food Project addresses these issues with an integrative model that allows young people to develop communication, teamwork, and leadership skills, find meaningful employment, and make a connection to the land and to the natural environment that will stay with them for a lifetime. Furthermore, by engaging youth in social change, environmental protection, and the food system movement, we build the next generation of civic leaders and practitioners.

Ahead of us are exciting challenges and opportunities. We are committed to expanding our local youth programs and our impact on Boston's metropolitan food system, while creating materials and workshops for those who wish to create similar programs around the country. This manual is one of many publications we created to share our work with those committed to a similar vision.

Patricia Gray

Exective Director

The Food Project, Inc.

Acknowledgements

The farmers' market is a critical piece of The Food Project's broad vision to create a local food system in Boston. Yet, the market is not just about the exchange of vegetables. It is a place where a community gathers to share and celebrate both the fruits of city gardens and the gardeners that produce them. It is a place to come together – to talk, laugh, and linger. Historically, societies that have engaged in vibrant trading of crops in market places have been more likely to delve into the broader fruits of cultural exchange – the sharing of languages, culture, religion, literature and world views. In years past, while fruits and vegetables were being thrown from ships to wheelbarrows to be carted to homes, agreements were formed, wars waged, ideas exchanged, friendships made, minds stretched and perspectives changed forever. Societies were and are changed by the interactions that happen in the market place – especially when these conversations are sparked by thoughtful and committed individuals with differing perspectives but common visions. It is in this age-old market tradition, intermingled with our modern vision to create personal and social change through sustainable agriculture, that The Food Project runs farmers' markets.

The writing of this manual has been as much a community endeavor as the market itself. First of all, many staff, gardeners, and neighbors have worked together to create a market worth writing about. The manual itself has changed as The Food Project has gained greater insight into the various aspects of market management. It is with the amalgamation of all the lessons and insights of the past that we offer this manual as a guide both for our own staff and for other markets throughout the country. Many thanks to Colleen O'Brien and Sara Coblyn for the first two drafts of the manual and for their diligence in creating a quality and community-oriented market. Thanks also to The Food Project farmers who have provided beautiful produce for the Dudley Town Common market over the years: Don Zasada, Martha Boyd, Carlos Gonzalez, Gideon Porth, Amanda Cather, and Danielle Andrews. A special appreciation goes to our neighbor gardeners who have sold their vegetables at the market: Sylvia Gaynor, Honoria Correia, and Joe Freitas. Other farms have also helped guide our market to success as well: Shartner Farm, Drumlin Farm, and Re-Vision House. The support from Liza Veras at DSNI has been invaluable. Thanks, too, to the residents of the Dudley neighborhood, to our local businesses and organizations, and to the Parks and Recreation Department.

The mission of The Food Project's market is to provide organic vegetables at an affordable price to the community. Without financial support from our network of foundations and corporate sponsors, the market could not exist. This manual is made possible through funding from the W.K. Kellogg Foundation.

Finally, a resounding applause for the youth that have, year after year, brought great energy, patience, and care to the Dudley Town Common marketplace. Their investment in the food they have grown and the interest they demonstrate for the residents of the city create an atmosphere that welcomes and celebrates. It is this exuberance that births a magical place for food and fellowship.

Kristin Brennan

Dorchester, Massachusetts

April, 2002

Additional Resources from The Food Project

French Fries and the Food System:
A Year-Round Curriculum Connecting Youth with
Farming and Food—From Seed to Market to Table

This agricultural curriculum features powerful, original lessons written and developed by The Food Project's growers and educators. Organized by season, the material teaches youth how to develop a deep understanding of and appreciation for the land and for local food systems. Personal, first-hand stories of learning in the field complement each lesson and encourage further exploration. Lessons can be done both indoors and outside and can be easily adapted by instructors working in school-based plots, urban food lots, community gardens, rural farms, and environmental education programs.

Growing Together: A Guide for Building Inspired, Diverse and Productive Youth Communities

This resource book is designed for communities of all ages andcan be used with many different organizations. Written as a comprehensive, practical and lively guide, it shares The Food Project's three-part model that encourages all members of a community to grow together through meaningful work, shared standards, and interactive learning. The book describes the role of meaningful work within communities, outlines a complete process of establishing and maintaining shared standards within a community, offers over 100 exercises that bring learning, reflection and energy to any program, provides tips for facilitating groups, processing activities and building inclusion, and includes rich photographs and inspiring stories to complement the text.

Program Manuals

These manuals describe the nuts and bolts of running all areas of The Food Project, including: the Summer Program, the Academic Year Program, the Volunteer Program, the Alumni Program, Farmers' Markets, Rural Agriculture, and Urban Agriculture. Together, these resources describe in detail the implementation and program management of The Food Project. These manuals will assist those who want to develop similar work in their own communities.

D.I. R. T. : The Next Generation

This video is the story of a diverse group of teenagers who break through their stereotypes about one another to become a close-knit community learning leadership, public speaking and farming skills. The 22-minute video is a glimpse into the spirit of The Food Project from the eyes, words and voices of the young people who have experienced the program. An ideal way to learn more about The Food Project, this youth-produced video will also serve as a springboard for discussion about a model that is thoughtfully and creatively challenging youth to build a better future for themselves and their communities.

Other Products:

T-Shirts

Mugs

Baseball Hats

For information on ordering these or any other publications by The Food Project, please contact us at:

The Food Project, Inc.

Attn: Publications

P.O. Box 705

Lincoln MA 01773

Phone: 781-259-8621

Fax: 781-259-9659

e-mail: outreach@thefoodproject.org

www.thefoodproject.org

Table of Contents

List of Attachments

Introduction

- *Welcome*
- *History of the Farmers' Market*

Welcome to The Food Project Farmers' Market

By the time we bring the vegetables to the market, our knees are dirtied, our fingers smell like basil, the backs of our necks are darkened, and our hands are rough and callused. We don't feel tired in the same way we used to, in the late afternoon, after a day of fieldwork. It is nearing the end of summer. The adrenaline has lasted from our morning activity, when we pulled red peppers and eggplants gently from their plants, lifted skinny carrots from their soft soil bed, snapped collard green leaves from their hearty stems and gathered their huge leaves into bunches of ten, winding a sturdy rubber band around their base. Our vegetables come out of the box truck, stacked neatly in bright orange sunshine crates, heavy with the weight of beets, corn, and potatoes. Onto long tables, arranged in bushel baskets and steel tin buckets, the newly harvested produce is guided into various categories – greens, roots, brassicas, fruits and summer vegetables. Signs are raised, price boards are tied to the sides of the tents, and the cash box is opened for business. Soon, the bustling increases, the tables are surrounded by a crowd of people, and the sound emerging from the scene is that of a marketplace of old, overflowing with familiar and foreign languages, shouting out prices and product names, pointing and gesturing, requesting and thanking. The young people that work the market table respond to their customers with care and precision. Two months ago they walked onto the land without any gardening experience but now they have pulled weeds, spread compost, set up drip irrigation systems, and harvested vegetables. These vegetables, their vegetables, are what they offer to the community that surrounds the lively stand.

The Food Project has been participating in farmers' markets throughout the Boston area since the organization's founding in 1991. In 1996, we began our own farmers' market on the Dudley Town Common in Roxbury, MA. The vision of the market was to create a lively and colorful marketplace, where residents of the Boston area, regardless of income, could purchase affordable, culturally-appropriate and freshly-picked organic produce for themselves and their families. Throughout the years, the farmers' market has lived into this vision. Vegetables, grown on The Food Project's urban and suburban farms by teenagers from diverse communities, are harvested, washed, and transported to the farmers' market site. Residents of the city purchase the fresh, delicious vegetables, and then witness their money cycled back through the local economy through the purchase of seeds and soil amendments for the city land. Young people are at the heart of this activity - weighing, bagging, and serving customers. While building skills in math and customer service, the youth develop an understanding of the management and running of a small business. At the end of the day, the market is closed up and the surplus food is delivered to area shelters. In a larger sense, the Dudley Town Common Farmers' Market is one step along the way to the goal of a local food system where food is grown, harvested, sold, and served all within the Boston area.

One cannot think well, live well, sleep well, if one has not dined well.
Virginia Woolf

The History of The Food Project Farmers' Market

In 1992, The Food Project opened its first youth-run farmers' market in the South End, a neighborhood just south of downtown Boston. The youth who ran the market were from Lincoln, Roxbury, Mattapan, and Concord, and other towns in the metropolitan Boston area. It was a simple stand with vegetables displayed in wicker baskets, signs with brightly colored youth drawings of beets and carrots, a cash box, a blue tarp, and smiling young people from the city and suburbs of Boston. Through the farmers' market, The Food Project was participating in a street-side food business, where young people could get a snapshot of food distribution that differed from large supermarkets and malls. In addition, youth from the suburbs were able to get a taste of the life of the city.

Our market migrated between low-income communities around Boston where we increased access to local food while spreading the vision and mission of The Food Project. Yet, The Food Project was interested in finding a permanent home for the farmers' market. We wanted to serve a need as well, providing for a neighborhood that might have limited access to fresh, organic vegetables. A commitment to one community became a central goal of The Food Project's market.

From the beginning of the organization, The Food Project founder, Ward Cheney, explored ways to build a truly local food system, impacting both youth and adults from different backgrounds, cultures and economic situations. The process of creating a local system is multifaceted and The Food Project knew that it was intimately dependent on a close connection to a particular community. The Food Project identified the Dudley Street Neighborhood in Roxbury and

North Dorchester, MA as a place that held many of important features that we valued. First of all, the neighborhood had a strong grassroots community development organization, Dudley Street Neighborhood Initiative (DSNI) that was supportive of The Food Project's participation. The area was a low-income area with limited access to local, organic fruits and vegetables and had a high rate of unemployment, poverty and environmental and social degradation. The neighborhood housed a large proportion of open and abandoned land that held the potential for growing food. In addition, the neighborhood had a history of agriculture and was already spotted with gardens, tended by immigrant residents who had rich agricultural backgrounds from around the world.

After committing to the neighborhood, The Food Project began to hire young people from Dudley. At the same time, we identified vacant land that could serve as a site for food production once it was cleaned and remediated from contaminants. We found a location for a farmers' market at the Dudley Town Common, a Parks and Recreation space located at a major intersection for walking, busing, and car commuting (See Attachment 1: Directions to the Market). Throughout this time, we cultivated partnerships with community organizations with similar interests and ideas about community and economic development, which have been instrumental in communicating the value of an open-air market to residents of Dudley. By 1995, we were hiring 30% of our youth from the Dudley Neighborhood. We had cleaned up and grown vegetables on a half-acre vacant lot on Langdon Street and solidified our permit for a permanent market site at the Dudley Town Common. The solid relationships we had built with DSNI and Nuestra Communidad, community development organizations interested in building a healthy economic, social and environmental atmosphere for their residents, helped us in the early stages of success with the market.

Attachment 1

Stabilizing our location in the Dudley community generated an incredible amount of momentum for our farmers' market. Since 1996, we have held the market at a consistent time and day. Customers rely on our presence and know that rain or shine, The Food Project Market will take place. Over the years, other vendors have attended the market—farmers and gardeners who have provided both character and fresh products each week. Bob Schartner, a vegetable and fruit grower attended the market for four years, bringing fresh corn, peaches and apples to excited customers each week from July to November. Sylvia Gaynor, a Roxbury resident and gardener, has sold a special variety of calaloo from her native Jamaica for over five years. Other growers (Drumlin Farm, Re-Vision House, Joe Freitas, and Honorio Correia) from both the city and the suburbs, have come and gone over the years.

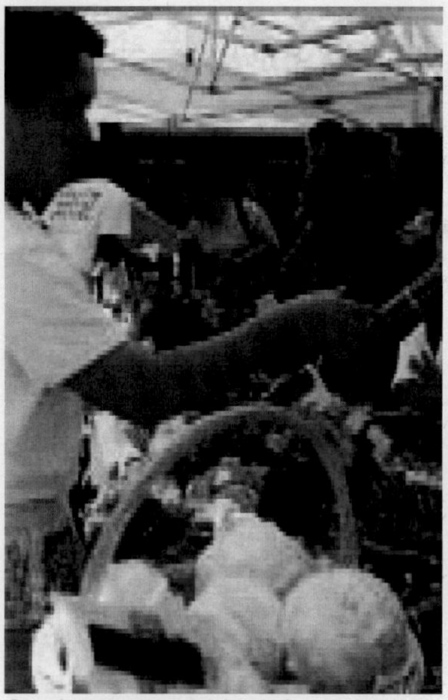

Good customer service is key to a successful market.

In 1999, we began running an additional market on Tuesday in Codman Square, Dorchester, but soon realized that our youth felt less connected to this neighborhood because of its distance from Dudley. Still wanting to increase access to local, organic vegetables, we decided to run our Dudley Town Common market twice a week. We were initially concerned that our revenue would stay the same, simply splitting the current customer base in half. However, we improved product sales at both markets with the added market. In addition, we provided our young people with more opportunities to run the stand throughout the season.

From the beginning, The Food Project wanted to create as diverse and full a market as possible for our customers. Over the years, we have sold organic gardening supplies, prepared foods, bread, honey, and eggs at the market. We held a chef demonstration each Thursday and invited several organizations to hand out materials, share important health information, and entertain our customers. We ran a farmers' market bus campaign that communicated the dates, locations and times of our market, as well as celebrated the idea that the food was grown locally and using sustainable practices, and planted and harvested by youth. We developed bright and colorful banners, and bought new tents and tables. We offered over 35 different kinds of vegetables at the market, selling calabaza squash, collard greens, okra, artichokes, and specialty hot peppers. We grossed nearly $19,000 at the markets in 2001 and distributed over 35,000 pounds to customers who used cash or coupons they had to purchase the freshest food in the area. We are proud of our market today because of the people who have carried it along its journey to the Dudley Town Common. Whether you are reading this manual to begin your own farmers' market, to improve the one in which you already participate, or to continue the tradition of running the market at The Food Project, you are entering an exciting time in the history of open air markets and The Food Project's market.

Planning

- *Create Your Plan*
- *Market Permits*
- *Equipment and Supplies*
- *Finding a New Market Site*

Create Your Plan

A detailed plan for a farmers' market is an invaluable tool for guiding a successful and smooth market season. When establishing a new market, it is important to consider the methods and strategies of well-run farmers' market to guide your own plan. There are several books and resources available that offer practical steps and insight into running a tight and smooth market. Speaking with market managers, farmers, and market customers is valuable way to guarantee that all details have been accounted for in your overall plan. If your market has existed for several seasons or more, it is important to use your own data to inform your market plan. Keeping records, notes, and observations year to year creates useful resources for determining the best aspects of your market and the elements that need to be worked on.

At The Food Project, prior to the market season (specifically December and January), the Market Manager reviews the farmers' market analysis, sales records, quarterly reports from the previous season, and current annual plan before setting out to achieve the goals for the market season. The Food Project Farmers' Market Analysis is produced by the youth during the fall and early winter and consists of a six-part curriculum. The results produce both qualitative and quantitative results, including customer survey feedback, the salability of particular vegetables, and the business skill development of youth. The analysis report holds information on customer preferences and trends over time in market revenue, vegetable popularity, and size of customer base. Throughout the season, the Farmers' Market Manager keeps detailed sales records and these numbers can provide a guide for vegetable needs and cash expectations. Quarterly Reports written by past managers and Annual Plans exist both on computer and hard files in management records and these provide an overview of market goals, successes, and challenges.

Once the manager has obtained these documents and reviewed them thoroughly, it is time to meet with the urban grower, the urban educators, and the program director to make decisions on start and end dates, educational initiatives, events, vegetable varieties for sale at the market, and youth training. The next step is bringing the initial decisions to the Program and Production Team (growers, directors and youth development staff), who will finalize the dates and logistics in the context of the whole program.

Choosing the Dates of Operation

Deciding when to begin and end your farmers' market can be challenging. The decision requires a sensitivity to spring and fall weather, the knowledge of the growing season of various crops, a sense of when farmers would be ready to sell their produce, and a sense of the market potential during the times when less produce is available. Will your customer base keep coming for spinach, lettuces, and radishes when there is no corn or tomatoes? You may want to delay the start date, if your production plan includes only a few early greens and beets, for example. Are the root vegetables available in the fall popular enough to make it worth your time and energy to keep it open through November?

At The Food Project, and in many spots in New England, we typically do not start farmers' markets until June, unless season extension is used or transplants are a significant part of the market merchandise. Since 2000, The Food Project has begun its market during the first week of June since our urban grower built hoophouses that expanded the early vegetable availability. On the other side, the hoophouses and the warmer climate in the inner city of Boston allowed us to extend the market until the beginning of November. Making decisions with both the growers and youth program staff allows you to choose market dates depending on the diversity of the harvest and on youth availability (to run the market). The urban grower is responsible for the production and distribution of vegetables grown in the city and the rural grower is responsible for suburban production and distribution, and both growers can provide essential information to your decision-making process. Critical questions to ask are: What are the growers' plans for season extension? What number and variety of vegetables makes for a diverse market? Do you plan to sell transplants and if so, when will they be ready?

Because of the relative warmth in the city, the urban lots provide all the produce for the early June markets. By the third week in June, the suburban farm begins to deliver produce to the market and continues to do so throughout the summer and fall. A June market may offer 5-7 different types of vegetables, including lettuce, salad greens, carrots, beets, spinach, peas, and strawberries. Transplants are popular at early spring markets because there are many gardeners in the area. Talk with the Lincoln grower about offering transplants to sell at the market. Extra transplants may be available or the manager could gather the materials to start plants. Ask the grower in January for advice and guidance about preparing transplants for the market.

Value-added products produced by The Food Project Kitchen, other growers, or food artisans can supplement an early or late market table. These products could include bread, baked goods, salsas, honey or greenhouse tomatoes. In order to fill out the spring tables with a diversity of products, talk to The Food Project's kitchen manager about ideas for items that could come from our kitchen. In addition, contact people we have worked with in the past. Verrill Farm, Haley House, and other local food artisans have wholesaled items to The Food Project in the early spring and late fall, and sometimes throughout the season. Don't be afraid to explore new ideas and products based on your review of data and surveys. When considering new options, however, keep in mind that additional products require planning and possible pick-up and distribution time.

Create an Outreach Plan

The farmers' market is full of hustle and bustle on a summer day.

An outreach plan is essential to the success of any market. Traditional forms of advertising are not always the most effective for generating business since farmers' markets are often attended as the result of personal and informal word of mouth modes of communication. The Food Project uses a variety of avenues to get the word out about our market. The Farmers' Market Manager is encouraged to take advantage of these established outreach paths as well as be creative about generating interest in and business around the market.

As a first step to developing an outreach plan, we encourage you to read this manual closely, focusing on the sections on marketing and advertising. Use the sections to spend some time thinking about the upcoming season. Draft a plan to communicate the value of the market to consumers. Make a list of the ways you might like to explore innovative outreach approaches. Contact representatives from the local community about meetings or events where you can give presentations about the market. Ideas include:

- Conduct a presentation to the staff of a local WIC program (Women, Infants and Children coupon program) or nutrition center on how to attract their constituents to the market. Many of these programs are housed in community action centers or health centers.

- Visit a local school and do an interactive presentation with students about the market. Hand out information about the market that they can take home to their parents. In the past, The Food Project has given coupons to kids and scheduled events at the market to increase customer attendance.

- Plan a certain number of celebrations throughout the market season. Hold one at the first market and one at the last market. Highlight different vegetables throughout the season with samples of different varieties and recipes.

- Deliver flyers to local community centers, churches, offices and clubs in the area and talk to staff about the market.

Hours of Operation

Choosing the hours of operation for a market is crucial. If the majority of the customer base will be composed of commuters or people on 9-5 working schedules, then the market needs to accommodate those who are working during the day. At The Food Project, the market has historically opened at 4:00 P.M. and closed at 7:00 P.M. We have found that this time frame fits both the schedules for our youth development programs and the shopping habits of our customers. We are strict with our opening and closing times. This highly structured approach benefits both the youth and our customers. Our young people respond well to the consistent structure of the market activity because it allows them to be intimately involved in the processes of setting up and taking down the market site. Consistent time frames and expectations create an atmosphere where the youth can excel at market tasks. In addition, our customers come to depend on us honoring the times we say we are there.

There are some time adjustments due to the changing of the seasons. As the autumn darkness comes earlier in the day, we often open the market by 3:30 P.M., and close by 6:30 P.M. This ensures the safety of our staff and youth. As these times change, it is important to inform customers (using flyers and signs) before you shift the market hours of operation, so that they can plan accordingly.

Market Permits

Permitting requirements for farmers' markets differ from site to site. Depending on where your market is being held, there may be a site permit required. In addition, depending on what products you will sell at the market, it is also necessary to investigate food permits. As a Market Manager, you are running a small food business that requires attention to detail. While selling raw produce at a Farmers' Market is probably the most benign type of food business in terms of health and safety regulations, make sure that the running of the stand follows all pertinent health policies. It is important to be aware of the rules governing farmers' markets and stay updated as they may change. Resources for Farmers' Markets are found in the Department of Food and Agriculture and the Department of Public Health.

The Food Project's current market site is at the Dudley Town Common.

Site Permit

The Food Project Market is held on public park land and requires a permit from the City of Boston, Parks and Recreation Department. In our case, the decision is made both by the city and by DSNI (Dudley Street Neighborhood Initiative), a neighborhood organization that holds eminent domain rights for the triangle area that includes the Dudley Town Common. To account for possible delays during the various steps of permitting, it is important to start the process by the end of April or sooner. The steps for obtaining the permit are as follows:

- Request a letter of support for use of the site from DSNI. The contact person at DSNI for park permits then sends a letter directly to the Commissioner at the Parks and Recreation Department.

- Send a letter of request to the Parks and Recreation Department for use of the market site. Address the letter to the Commissioner (See Attachment 2: Sample Permit Request.).

Attachment 2

- A formal application for site use will be sent to you. The application will request that you submit proof of insurance for staff, youth, and the site. Obtain a copy of insurance from the Office Manager and submit it with your application. In addition, hold on to a copy of insurance to be held with you at the site at all times.

- Call to follow up on your application, making sure that all the materials were received and that the process is moving along.

- Once you have obtained your permit, laminate it, and place it in a safe, but accessible container that is brought to and from the market site. It is imperative that the permit is with you at all times when the market is open.

Parking Permits

Parking availability is an important resource for widening the customer base of any farmers' market, but obtaining parking privileges in an urban area is often a difficult task. The Food Project chose the Dudley Town Common site in Roxbury because it was a major intersection with several bus interchanges and a lot of foot traffic. Therefore, our market is not dependent on our ability to accommodate customers with cars. However, Dudley Street and Blue Hill Avenue are commuter routes for many people driving to and from Boston for work. Knowing that our customer base could be expanded by trying to attract commuting drivers, we have investigated various parking options for our market customers.

Attachment 3

We have made several requests in the past to the Department of Transportation for permission to use public parking lots during the hours of our market operation (See Attachment 3: Sample Parking Request) . However, permits from the City of Boston are difficult to obtain because of the high demand for public parking in the business storefront area. Another option is to request parking privileges from private organizations and businesses in the area. In the past we have requested use of the parking lot at St. Patrick's Church (across from the market site), but to date, our efforts have not been successful. The Market Manager should continue to explore options like this as well as others. Be Our Guest Catering Company, local schools, and other city-owned lots may provide some public use of their lots. Market coupons are a great way to barter use of privately owned space.

Scale Inspection

Accurate, consistent, and functioning scales are essential equipment for a farmers' market. The City of Boston, Inspectional Services Department, insures scales annually for all types of businesses, including farmers' markets. Before inspection, it is important to repair old scales and buy new ones if needed. Due to constant handling, there is a lot of wear and tear on farmers' market scales. The most important thing, however, is that the scale is functioning and weighing accurately. Therefore, a scale with broken glass can still be inspected and used for the season. With that said, due to the cost of the inspection approval for each scale, limit the scales to what is needed for that season.

DIRT Crew Member Alex Gomes weighing produce.

Schedule an appointment to have your scales licensed with the Inspector (to be found in the 1010 Massachusetts Avenue complex). Inquire about the cost per scale, since the inspection sticker fee is likely to increase over time. Often, the Inspector will come to The Food Project office or farm lot but be ready to bring the scales to 1010 Massachusetts Avenue if necessary.

Vendors' Permits

The bigger and more diverse a farmers' market, the more attractive it will be to the public. Multiple vendors offering a variety of products is an effective strategy for increasing the business and vitality of a market. Having uniform and clear guidelines and agreements for the vendors guarantees professionalism and efficiency at your market. Necessary permits are part of the process of assuring commitment and value. If you are entering into an already established farmers' market, then the market manager will inform you of the requirements for you to sell your product. However, if you are managing a new farmers' market, or continuing to manage an established market, it is important to have guidelines and agreements for each of the vendors.

At The Food Project market, all vendors must have proof of insurance for their own businesses or farms. There may be exceptions to be made for neighborhood gardeners who sell their own produce by our side but any other food product must be made in an inspected commercial kitchen. Copies of food permits and insurance for each of the vendors should be kept in the manager's files to be brought out upon the request of an inspector. We also require vendors to complete an application to sell at our market (See Attachment 15: Vendor Application and Guidelines) and pay a small fee. **Attachment 15**

Temporary Food Services Permit

Fresh, uncut vegetables and fruits do not need a permit to be sold at a farmers' market. However, all other foods require permits to sell. The Division of Health Inspections requires a temporary food service application for any food that is prepared on-site or off-site in a commercial or licensed residential kitchen. Requirements for open air, prepared food distribution, may change each year. If a new product is being researched to sell at the market, it is important to first look at the permitting implications. In the past, we have obtained a permit from the Board of Health that allowed us to sell fresh bread and honey. With the addition of The Food Project's commercial kitchen and the production of various foods and food products, the Market Manager is responsible for ensuring permitting for all of the various items to be included on the market table during the season.

Keeping track of sales and customers is a challenging task during a busy market.

In preparation for the recruiting or creating products for market, review the history of The Food Project's research into permits found in Farmers' Market files. Request updates on this information from the Department of Food and Agriculture, or the Division of Food and Drug at the Local Board of Health. Refer to the files on permitting history. Call other farmers' market managers or the Federation of Massachusetts Farmers' Markets to update and continue the research process. The Board of Health will give full information on updated health regulations.

Other Permits

Farmers' Markets can expand in many different ways. Depending on new initiatives that may develop within the organization, the market manager should be ready to research any other permits required to ensure the success of additional activities and products.

Equipment and Supplies

Equipment is an important part of any farmers' market. Because the market is a temporary stand, and depends on materials that are handled often, it is important to care for the Farmers' Market equipment properly and replace it when necessary. When choosing new equipment, keep in mind the importance of mobility, durability, and functionality for the market.

Attachment 4

At The Food Project, the market manager keeps an inventory spreadsheet (see Attachment 4: Equipment Inventory Spreadsheet). The spreadsheet is the master list of all the equipment for the farmers' market stand. The list includes information on the number of items, the optimal number needed, the condition of each of the materials, and maintenance recommendations. An inventory list helps the manager to keep on top of all the equipment and replace materials when needed. Consult the previous year's inventory list during early

March. Some equipment will need to be repaired and some replaced. Allow time for ordering and repairing and keep in mind that farmers' markets throughout the area will be doing inventory and ordering during the same months and might delay or limit items that you need.

Some things to consider when evaluating your equipment needs:

- Budget. Check the budget for Equipment and Supplies allocations before you order anything new.

- Inventory. Check the actual supplies in the shed for what you can fix and what should be replaced.

- Donations. Many organizations or businesses are interested in donating to causes they are interested in. Review the following section on donations to brainstorm on sources for some of your materials and equipment for the season.

- Suppliers. Consult the ORGANIZATIONS database in the computer for the Farmers' Market vendors. In the notes field of the database, do a find for "FM Supplier" and sort through records for the appropriate supplier.

- Consumable items. Consumable items can quickly drop to critically low numbers, so keep track of them. These include pens, receipt pads, markers, produce bags, extra paper for signs, and chalk. When chefs visit and cook at the market, it is important to replenish plates, cups, forks and knives during the season.

- Tools. Make sure you have the necessary tools to fix equipment that may break during the season. Tools can also include duct tape, tent parts, and bungee cords.

- Final Inventory. At the end of the market season, inventory all market items and revise the inventory spreadsheet accordingly. This should also include a general budget estimate for repair or replacement of broken equipment.

Donations

Donations are an effective way to keep supplies up and costs down for a farmers' market. It is also a way to include community organizations, businesses, and individuals in the market. Many local people and businesses are interested and happy to help a worthy cause and they only need to be approached with a request.

At The Food Project, we rely heavily on donations for consumable as well as non-consumable items in order to keep our market costs down. The market manager should be guided by the funding team on how to request and track in-kind donations. Ask other staff members in the organization for suggestions about requesting donations. All in all, the most effective approach is to be direct and clear in your request. Make sure you have the specifics about what you want and need. Have details on hand including numbers, type, style, and size. It is best to overestimate as far as number, as there can always be overstocking for the next season. Be sure to document each donation in the database, so that other staff will be informed that a certain business has been contacted for donations. In addition, send a thank you note expressing our gratitude and recognizing the importance of local support for the work we do.

There are many small and larger items needed for the market that could be easily spared by businesses while significantly reducing market costs. Be creative and active in seeking out donations. Over the years, The Food Project has benefited greatly through generous donations. We have received discounts from EZ Up Tent Company, which has allowed us to improve the infrastructure of our market. Local and national supermarket chains are often willing to donate

Attachment 5

plastic bags (produce and handle) to our market (see Attachment 5: Sample Bag Request). The produce managers at Bread & Circus and the Stop & Shops in Jamaica Plain and Dorchester have all donated bags to our market. Products can also be donated. We received a donation of ten cases of organic macaroni and cheese from Annie's Homegrown of Vermont to sell at our market. Not only did the macaroni and cheese bring in some extra revenue, it was very popular among customers and exposed people to a healthy, local, organic option for a quick dinner or lunch.

Finding a New Market Site

Locating and establishing a new market at an additional site is not an easy task. There are many elements and details to consider and often these considerations are complex and multifaceted. Fortunately, The Food Project has only needed to do this on a few occasions because it can be a drain on money, staff, and resources. But we realize that in the evolution of any business, a new market site may become necessary. The action to relocate a market or start a new one is often made during the annual planning process and by the whole staff. Review your annual plan to determine the plan for locating your markets.

If you are starting a new market, location should be of primary concern. The location of your site can be the make or break of your success as a business. Consider certain factors for site location:

- Visibility
- Foot traffic, public and private parking (preferably free) and public transportation characteristics
- Other fresh-food related competition in the area
- Community's desire for open, public market
- Economics of residents and employees in the area
- Presence of peer organizations who can actively support, promote, or publicly celebrate your efforts
- Aesthetics of the site
- Convenience to program and to customers
- Adherence to The Food Project's mission

Trying out a new market site.

It is best to investigate the characteristics of your proposed market site by talking with people in that area of town. Ask yourself and other staff members, concerned citizens, and leaders in peer organizations the following questions:

- Can you see the market from the bus or from office windows?
- Is it tucked away or on the street side?
- Do potential customers live and work in the area?
- Do potential customers stop at the site to change buses or pass through your area on their way home or to work?
- Is it convenient for them to stop in their cars or shop and then hop on the next bus?
- What other organizations provide a similar product to residents and area employees?
- Will you be in competition? Are you a niche market for these items?
- What kind of relationship do you want to have with area businesses including those that you are not in direct competition with?
- Are your customers higher income or lower income or is there a mix?
- Is there a business association you can join to legitimize your presence in this new place?

The Food Project is a non-profit organization with a social mission (see the first pages of the manual). Our primary commitment at the markets is to make healthy, affordable, and organic produce available to our neighbors and constituents who need it. Although we do seek to cover as many of our costs as possible, our markets are not driven by profit. We seek to serve people according to our vision. Therefore, when considering an addition site location for a market, refer to The Food Project mission to identify how it fulfills our mission.

Once you have decided on a potential location and purpose of a new market, go to leaders and public officials in the area to assess the logistics of introducing a farmers' market in that community. When starting a market, it is crucial to gain the support of many entities and individuals within that community. These supporters will help to make marketing strategies and outreach ideas a reality. It is also likely that these organizations may become potential funders or financial supporters.

Site Proposals

In 1999, a search was done to locate an additional market site in Roxbury and Dorchester. If you are attempting to locate another site, review any history of this type so as not to redouble that effort (see

Attachment 6

Attachment 6: Sample Site Use Proposal**)**. It might make sense to call some of the places already approached to see if their availability has changed.

Steps in developing a site proposal:

- Identify who has ownership of the site you are interested in.
- Identify key community organizations, local elected officials, and individuals in the community who you could name as supporters and approach them with the idea of the market.
- Assess the needs of this community for a farmers' market.
- If your assessment identifies a need, craft a proposal for the benefits this market will bring to the community.
- Gather support letters and detail the specifics of the needs for the market including site use, electricity, parking needs, space needs, and so on. The details of this proposal will depend on the needs of the particular organization to which you are applying. They may also ask you to create a map that shows how you will use the area for the market in order to see if it will impinge on pedestrian or other use.

Standing proudly behind their market set up, a crew from the Summer Youth Program is ready to receive the day's customers.

- When dealing with a city entity or bureaucracy, follow up is crucial. Be prepared to make several calls without getting any calls back. Your persistence will be the key to your success.

Once you have obtained a new market site, gather information on your potential customers. You may have already gleaned some information through your research in obtaining the site. It is helpful to know the ethnic and income breakdowns for the area surrounding the market site. Market surveys are beneficial to determine the types of products the community would be interested in seeing at the market. You may be able to work with a community- based organization to get this done. The surveys can serve as a marketing tool. Try partnering with a couple of organizations that expressed support for the market and hold a community wide meeting to inform residents about the upcoming market and get their feedback.

Kristin Brennan, former Farmers' Market Manager in the middle of a typical market.

Outreach and Education

- *The Overall Marketing Plan*
- *State and Federal Coupon Programs*
- *Vendor Recruitment*
- *Neighbor Gardeners*
- *Special Community Outreach Programs*

The Overall Marketing Plan

A market's success from year to year is due to a great extent on how well its value is communicated to the outside world, and especially to target customers. Through designing a diverse and organized marketing plan, which can be based on the analysis of service and sales from previous seasons, the market can respond directly to the community's needs and desires.

Produce for the farmers' market comes from city lots and the farm in Lincoln, MA.

Every year at The Food Project, we develop an extensive plan for letting others know about the resource of health, food, and education that the market provides. Part of the success we have found in integrating young people into the market comes from the fact that the market is a legitimate business and the goal is to please the customers with a quality product that they see as desirable. Knowing the importance of the work they are doing, the youth respond by giving high quality customer service and the customers trust the inherent value of the product.

Understanding the following concepts will help you design an effective communication plan for your farmers' market:

1. Know and Trust Your Product

Before communicating the true value of the market to a potential customer base, it is important to know and trust your product. The produce delivered from the farms to the markets each week is a trusted, quality product. The Food Project's growers are well-informed about the needs and desires of the community and each grower is skilled at growing the produce requested by the customers. When selling, it is important that the manager, the youth, and other staff are fully aware of the names and culinary uses of all the products at the market. Knowing in advance what will be available encourages loyal customers to return for certain items. You will feel more capable as a marketer if you can provide customers with a reason to shop week after week.

2. Understand Niche Marketing

There are many reasons why a person might visit and shop at a certain farmers' market – convenience, price, variety, organics, sustainability, and availability. One way to attract customers is through a niche product, something hard to find but popular among people who live in the area. Because the Dudley Town Common Market is located in a culturally diverse community, The Food Project has the opportunity to offer niche market products that customers may not be able to find in the grocery store. Another aspect of the niche product that a farmers' market offers is the aspect of "fresh" or "organic" which is not readily or abundantly available in urban supermarkets. A survey held year to year on customers' product wants can offer a helpful guide for market managers and growers. At The Food Project, the end of the season annual Farmers' Market Analysis offers some insight into the desires of the customers, but it might not help you to gauge the desires of the people who are not attending the market. Obtaining the input of non-customers can be done through a neighborhood door to door survey, speaking to a local nutritionist, or visiting a health center that may indicate what kinds of products are desired by the community but are not found in other local markets.

3. Location as Key

As has already been discussed, location for a farmers' market is key to its success. Having a visible site for passersby to stop and investigate the market is a significant marketing strategy. The longevity of a market's existence also ends up being its own outreach method. The strength of the Dudley Town Common Market is that it has been situated in the same place for five years and customers have come to rely on its presence there each year. Dudley Town Common, a park space situated at the intersection of five streets, is a place where folks come to shop for their stereos, attend church meetings, eat pizza and subs, and change bus lines. Because the market is at the hub of all of this activity, we are sure that many people learn about the market simply by passing it.

4. Communication with Your Customer Base

Developing a solid communication plan may be the key to reaching customers who are not currently shopping at the market. Communication should also be geared to remind customers who have shopped at the market in the past, but have forgotten when the season begins, or the kinds of products they can expect to buy. As you design methods for targeting potential customers, be cognizant of the following consumer attitudes and behaviors:

- Most of your customers will come from within a three-mile radius of the market.
- People need to be reminded at least three times of the market, whether it be through verbal or visual cues.
- Each potential customer receives relevant information in their own way, determined by the activity and events of their days. The key to reaching customers is in identifying these life patterns and announcing the market in places and in forms of communication that people are accustomed to. The best bet is to employ a variety of methods, places, and forms.

A customer buying fresh produce for that night's dinner.

5. Market the Mission of the Organization

The Food Project Farmers' Market is not only a business. It is also a training ground for youth focused around service, community involvement, local food systems, and job preparation. The market is a place to share with others the value of buying and eating locally grown organic produce. The farmers' market can offer a model for how an urban community that has suffered environmental and social degradation due to poverty, misuse of resources, and neglect from the city government can celebrate the local resources of people, food, health, and land. The market manager's task is to communicate The Food Project's mission in every way possible – to the youth, the community, and the customers. Accepting federal and state coupons is one way to bring customers into your mission but there are many other ways that the market demonstrates the values and vision of a vibrant, local food system made up of healthy, nutritious farm goods.

Methods of Communicating with Potential Customers

There are various ways to reach out to potential farmers' market customers. Be creative in your approach and methods. When preparing your marketing plan, take into account that the message and the messenger are equally important. Word of mouth can be more effective than even the most polished marketing schemes, so consider developing networks of customers and neighbors to spread the word to family, friends, and community members through your loyal supporters. Remember that there is a wealth of free advertisement available, especially for non-profits. To guide your process, refer to the past year's Market Analysis to guide your outreach and marketing strategies.

Consider the following advertisement methods used by The Food Project in the past as you design and implement your advertisement plan. Expand this list as you proceed since it is by no means comprehensive.

Local Organizations

The Dudley Street Neighborhood Initiative (DSNI)

The Food Project has benefited greatly through the peer relationship with the local neighborhood organization, Dudley Street Neighborhood Initiative. DSNI has an incredible ability to communicate with and mobilize residents for a cause. In addition to being very supportive of The Food Project's work with local teenagers, DSNI also views the market as a model for economic sustainability. In the past, The Food Project has advertised in several of DSNI's monthly publications. If you are interested in doing this, inquire to the publications coordinator at DSNI. At times, there has been a small fee for this service.

Merchants Associations

Merchants Associations exist to support the local economy. Mem- ber- ship in the local merchants association will provide you with resources and support as you continue to build and communicate the value of the farmers' market. Membership in the association legitimizes your business with other storefront operators, even though you are a seasonal farm stand. Associations are often connected to one another and membership in one would guarantee exposure to other groups throughout the Boston area. Begin by contacting all associations within a three to five square mile area and inform them about the market. If possible, attend merchant association meetings to share the benefits of the market with their members. The farmers' market has the potential to stimulate business in the area and be stimulated by others'efforts in advertising. Membership and information sharing in the business community will help to sustain the market in the long run.

Local Businesses

The Food Project relies on local businesses to provide us with the support we need to spread the word about the market. Each May, Food Project staff and youth request space in store windows for large posters that advertise the market. Many businesses agree to give counter space to our flyers as well, so that when people come in to eat or to buy, they can take an advertisement home with them. When talking with the businesses about advertising, we also use this time to check in with business owners to air any issues they may have with the market.

"We have great relationships with the businesses across the street from the market site. We often give them discounts on produce during the season and sometimes we donate extra produce, when it does not sell. We check in with business owners at the beginning and during the middle of the season to air any issues they may have with the market. The couple that owns Tim's Pizzas has occasionally bought produce (lettuce, peppers, and tomatoes) and sometimes gets a donation. On countless occasions, we have borrowed a tool or two from the mechanic shop, when a tent or a table breaks. We rely on these relationships to keep our business going and growing. I believe they appreciate our presence in the neighborhood. "

Young people from The Food Project conducting a workshop on one of the city lots.

Neighbors and Resident Groups

In Roxbury and Dorchester, residents live very close to each other. Many were born in different countries and therefore, different languages are spoken in the area. In several ways, the neighborhood is very active and involved in planning for their future. With this type of activity, often the primary method for spreading the news is by word of mouth. Door-to-door organizing is a great way to speak directly with a potential customer and get them to speak with each other about the food they buy each week at the market. Involve The Food Project youth in distributing flyers in the late spring and until the end of the harvest in October. (See Attachment 7: Trilingual Flyer, Attachment 8: Spanish Flyer, Attachment 9: Coupon Flyer). Youth find this energizing and understand how it fits into the bigger picture of the business. Because many residents do not speak or read English, it is essential that you find a young person who can translate the message to the Cape Verdean and Spanish speaking population in the neighborhood.

Attachment 7
Attachment 8
Attachment 9

Collaborate with other Food Project staff as the season nears, to see if there are plans to organize youth in other surveying projects in the neighborhood. Once you have a captured audience at the door, it is easy to share information about the market. We often hand out coupons to folks who have not heard about the market, as a way to encourage them to come and buy vegetables.

Community Organizations

Any one community may contain hundreds of formal and informal associations or groups that get together for a common purpose. The Food Project has an extensive database of peer organizations that may disseminate information about the market. Community Development Corporations (CDCs) keep lists of associations and may be able to supply you with contact information for the groups. You may want to form your own resident advisory group to plan the future of the market. We have organized such groups in the past but they have dissolved once the season ends. If you decide to go this route, remember to choose people who are leaders in the community and have energy to attend meetings when scheduled. Tasks for the group may include: organizing a market event, bringing several friends to the market each week, recruiting gardeners to sell at the market, or volunteering to translate for customers at the market.

Government Representatives

Every year the Department of Food, Agriculture and Markets publishes a list of farmers' markets in Boston and around Massachusetts. The Food Project Farmers' Market is typically listed as a result of our application to accept federal and state food coupons, but to ensure a proper listing, contact the director of market advertising at the department.

Legislators and local representatives love to learn about local events that positively impact the social, environmental, and economic atmosphere of a community. We have had incredible success in the past inviting politicians to the market and typically do so, in August and September, when youth involvement is high and the vegetables are big, colorful, and abundant.

Press and Other Media

Community Newspapers

Residents in Roxbury and Dorchester rely on community newspapers for reliable, relevant, and interesting local news. Be sure to coordinate your efforts with the Public Outreach Coordinator, so that you are not doubling efforts to advertise the market. Review previous advertisement efforts in Food Project files to assess their success and consult the Farmers' Market Analysis to plan for more appropriate advertisement efforts.

Community newspapers boast comprehensive calendar listings. The Dorchester News reliably posts Food Project advertisement in the "Community News" section of the paper. (See Attachment 10: Sample Calendar Announcement.) Other community newspapers have similar approaches to communicating local events. Although news reporters come and go, it is essential to find a champion at the newspapers, who knows the work of The Food Project and can promote us from within. In the least, know the name of the person who is responsible for posting the advertisements. When writing an announcement for the market or for other related events, include the location, brief

Attachment 10

directions to the site, days and hours of operation, and the products sold. Jazz up the advertisement by using words like fresh, convenient, and affordable. You want to be honest and at the same time, hook your potential customers.

Radio and Cable News

Communicate your interest in a radio address or cable news slot to The Food Project's Public Outreach Coordinator. They will advise you on the processes for communicating this way. Public Service Announcements (PSA) are effective ways to reach potential customers and youth with solid public speaking skills and/or language fluency may be able to communicate with customers difficult to reach through traditional, English-based news medium (i.e. Spanish or

Attachment 11

Cape Verdean programming). (See Attachment 11: Sample PSA.)

Food Project Efforts

Mailings

Reaching people through the mail can be an effective method of reaching former customers or people you know through other activities and connections. At The Food Project, as we continue to develop contacts in the Dorchester and Roxbury area, our database becomes an important resource for communicating the value of the market. Do a general mailing to neighborhood residents by zip code within a five square mile radius of the market site. Include market advertisement in other organizational mailings that go out to potential customers. These mailings may reach people who work in the area and can shop at the market before heading back to their own community.

The place to improve the world is first in one's own heart and head and hands, and then work outward from there.
Robert Pirsig

Food Project postcards (containing photos of young people holding vegetables or working in the fields) can provide great incentive for a resident to attend the market. A message that states the dates, times, and locations of the market may be enough to increase participation in the market.

The Food Project produces many publications and newsletters that can post an advertisement for the market. The communication committee, an internal Food Project working group, sets forth an annual timeline for organizational communication. Request attendance at a meeting early in the year to plan for space in an upcoming newsletter. Food Project mailings that are appropriate for market advertising include:

- Organizational newsletter
- Alumni newsletter
- CSA newsletter and information
- Annual appeal
- Mailings to volunteers
- Board correspondence

Consider hooking potential market customers through the issuance of a Food Project Farmers' Market coupon. You may decide to print the coupon in a newsletter or just distribute when surveying door-to-door. Consult your budget and decide how many coupons you would like to print for the season. Redemption has historically been low, but this may change as people learn more about the market.

The Food Project's Farmers' Market serves over 300 people each day it is open.

Banners and Sandwich Boards

Big, colorful, eye catching signs are essential for an outdoor market. The Food Project owns several sandwich boards and banners that are used on the day of the market. They can be found in the supply bins stored in the sheds. The signs are placed along both Blue Hill Avenue and Dudley Street and are effective for folks driving cars or riding the bus. The banner is usually hung up below the clock tower facing Boston. Design other signs for the market itself to draw people's attention to the display. These items may need maintenance from year to year, but will be long lasting, if taken care of by youth and staff. When you offer a special on one of the products at the market, let your customers know. Cardboard is hard to display unless you have a firm structure to post it on. Dry erase boards or chalkboards may serve the purpose of showcasing temporary sales.

Permanent signage is optimal for a farmers' market. At The Food Project, efforts have been made to establish permanent signage for the market site, which in Boston is a very difficult thing to do. Several years ago, there was a near success story for permanent market signage. The farmers' market manager established a relationship with the Department of Neighborhood Development (DND) to design and print banners for the area surrounding the market. For a variety of reasons, the pole banners were printed and never hung up. Continue to connect with DND about pole banners as you may find a champion in the department, who is willing to go the extra mile to make things happen. Be persistent with these efforts, as it is often just a matter of talking with a voice and not voicemail.

Paid Advertising

Word of mouth, on site signage, community newspapers, and radio spots will not hit everyone. If you have resources available to you, it is good to consider paid advertisements to reach a clientele that less prominent methods will not reach.

In 1998, the Environmental Protection Agency funded the Food Project to design and implement a public media campaign to discourage the use of pesticides in urban gardens. This effort included money to pay for advertisement about this issue, so that we could reach a broader audience with the message. We chose to post two color, bilingual signs on MBTA bus and transit lines in the area, encouraging the purchase of our fresh, local, organically grown vegetables. A significant amount of money was designated to design, print, and post these signs for six weeks. Although it is difficult to quantify the success of these efforts, we feel that purchasing advertisement space was worth the money, and in the least, complemented our other marketing efforts.

State and Federal Coupon Programs

"I will miss running the Farmers' Market because it is fun to stay late and hang out with your friends while helping out the community. It is satisfying to see that people can get locally and organically grown vegetables for a cheaper price. You know that you are really helping the less fortunate when you see people smiling because of all the things they can get with their Farmers' Market coupons and WIC coupons."

- Shatara Francisco, Summer Program Crew Member, 2001

Farmers' Market Coupons

The United States Department of Agriculture has been sponsoring the dispersal of farmers' market coupons as a way to boost the vitality of farmers' markets and to heighten food security in low-income neighborhoods. These coupons, given out to low-income mothers, elders in the community, and others, provide a critical resource for both farmers and families. The Farmers' Market Nutrition Program (FMNP) was established by Congress in July 1992 as a supplemental source of food income for the more general Women, Infants and Children's (WIC) Program, a federal program which allocates several million dollars each year to mothers who meet low-income guidelines, are pregnant, breast-feeding, and/or have children under the age of five.

The FMNP was passed by Congress to accomplish two main goals:

1. To provide fresh, nutritious, unprepared foods (such as fruits and vegetables) from farmers' markets to women, infants, and children who are nutritionally at risk.
2. To expand the awareness and use of farmers' markets by consumers.

In the neighborhood surrounding The Food Project's Farmers' Market, many residents meet federal guidelines for "low-income" and are eligible to receive services through state and federal food assistance programs. This assistance often comes in the form of nutrition education resources and services, health screening, and supplemental nutritious food. WIC (Women, Infants and Children) is a federal program that provides a range of nutritional services for families including nutrition education and breastfeeding support services. Their main program is distribution of WIC coupons which are provided to Massachusetts residents who are pregnant or have children under five years old and meet specified income guidelines. People who are accepted into the program receive several coupons (valued at $2.50 each) designated for the purchase of "Fresh Fruits and Vegetables." The WIC Program provides a directory of farmers' markets in Boston and tips for shopping at the markets.

The coupons are distributed through WIC Program offices or neighborhood health centers. There are specific coupons issued for farmers' markets and they should not be confused with the standard WIC supermarket coupons. Each coupon specifies which products can be purchased with them. In the case of farmers' market coupons, they are appropriate for the purchase of fruit, vegetables, herbs, and flowers. However, it is important to check the restrictions each year since they are bound to be slightly altered. Unlike the supermarket coupons, the farmers' market coupons do not need to be signed by the individual. It is important to note that change cannot be given for coupons.

The Food Project's Farmers' Market would not be able to serve the community as well without the WIC Farmers' Market Coupon program. Each year, approximately 60% of total sales at the market come in the form of WIC Coupons. Our collaborations with WIC centers are crucial because they represent our largest customer base. Connecting with WIC representatives in health centers is an effective method for reaching out to coupon recipients so that the can exchange their coupons for fresh, local vegetables. WIC Directors have allowed us to participate in WIC trainings and presentations for both the health center staff and the recipients. We have a comprehensive list of WIC

Directors in our database and examples of letters offering presentations and nutrition trainings to staff and recipients in market files. The market manager should focus on deepening these existing relationships by contacting WIC directors closest to our market by letter and phone reminding them of our presence and offering our services.

Attachment 12 (See Attachment 12: WIC Director Letter.)

Other Coupon Programs

"This new initiative encourages low-income seniors to improve their diets, while helping America's family farmers increase their markets for locally-grown fruits and vegetables." – Dan Glickman, Former USDA Secretary of Agriculture, 2001

The Department of Food and Agriculture (DFA) has two other coupon programs. One is the Seniors Farmers' Market Nutrition Program (SFMNP), which was piloted in 2001. The coupons were established under the USDA's Commodity Credit Corporation (CCC) and makes grants to States and Indian tribal governments to provide coupons to low-income seniors that may be exchanged for eligible foods at farmers' markets, roadside stands, and community supported agriculture programs. Like the WIC coupon program, the SFMNP seeks to:

1. provide resources in the form of fresh, nutritious, unprepared, locally grown fruits, vegetables, and herbs from farmers' markets to low-income seniors

2. increase domestic consumption of agricultural commodities by expanding or aiding in the expansion of farmers' markets

3. develop or aid in development of new and additional farmers' markets

Other coupons are directed toward low-income residents who are not women with young children or elders. These coupons can be picked up through local community centers. The elder and community center coupons do not hold the same restrictions as the WIC coupons but instead allow for a greater diversity of purchasing than the WIC coupons do, covering baked goods, honey, meats, and dairy in addition to fruits and vegetables.

Information about the coupon program can be found from the Department of Food and Agriculture, which publishes extensive information about farmers' markets on their web site and in their publications. At The Food Project, it is important for the market manager to have an understanding of the coupon programs since they make up such a large percentage of our revenue. Each year, the funding for the coupon programs comes up for renewal. Because of budgetary cuts for FY2002, the coupon program is being reduced. Since The Food Project receives over 60% of its revenue through the coupons, this funding reduction will be felt at the Dudley Town Common Market.

Farmers' Market Coupon Redemption and Outreach

In order to redeem the market coupons, you must become an approved market through the Federation of Massachusetts Farmers' Markets. The applications are distributed to markets in April, but you may want to call to verify the timeline with them, so that you are ready to redeem coupons beginning in June. The application is accompanied by a list of requirements for authorization to redeem all food assistance coupons. Read over the paperwork and identify updated guidelines. Information on coupon redemption is found later in this manual, under the Operations Section.

Young people do a variety of tasks--selling, stocking, bagging, and helping customers.

Below, there is a checklist of steps that suggest the way to conduct outreach to local WIC and community centers about the farmers' market. Contact the Massachusetts Department of Agriculture with questions and send away for an application for coupon redemption from the Federation of Massachusetts Farmers' Markets.

Here is a checklist of steps to take in your outreach to local WIC centers:

- By the end of April, craft a letter to WIC directors reminding them of the upcoming Farmers' Market season and offering any outreach services (see Attachment 12: WIC Director Letter). **Attachment 12**

- In the ORGANIZATION Database, in the "Our Organizations Function" field, locate all "FM Health Contacts." This will give you a working list of people who can be contacted in person, phone, or mailing.

- Send this letter to both WIC Directors as well as general health organizations notifying them of the coming market season. It may seem early in the season, but it is good to get people thinking about the market and whether they would like to schedule some presentations with our young people during the spring.

- Follow the letter up with phone calls introducing yourself and checking if the director has any interest in having us give a presentation to their staff or constituents or for any other collaboration.

- Farmers' Market coupons are distributed once a month, beginning in June and the final distribution is in August. The Department of Food and Agriculture will send you a notice letting you know when coupon distribution will begin. A few weeks before the first market send another mailing to "FM Health Contacts" which are listed in the database. It should include a letter announcing the dates and times of the market, between 50-100 flyers (1/3 of horizontal 8 1/2" by 11") to be included in the recipients' envelope, and several posters for a bulletin board in the health center (see Attachment 13: Follow Up WIC Letter and Attachment 14: Market Poster).

Attachment 13
Attachment 14

- You may want to do another mailing at the end of July to remind recipients and health center staff of the market calendar.

Note: Neighbor gardeners are not likely to be certified to receive farmers' market coupons and exchange them for cash with the DFA. Yet, many customers come prepared only to use their coupons for purchase. In the past, we have made agreements with neighbor gardeners that we will give them the cash value of the coupons that they collect. Then we simply redeem their coupons along with our own. It does not cost us anything to add their coupons to our stack and send them in. Yet, the exchange enables them to offer their produce to anyone who comes to the market.

Vendor Recruitment

Lively, full and diverse farmers' markets have the best chance of attracting a steady stream of customers. Farmers' markets in the northeast cannot easily match the one-stop shopping that is so appealing to people about supermarkets, but we can continue to provide fresher and affordable products to consumers. Managing a farmers' market means recruiting vegetable farmers, fruit growers, food artisans, or other product stands to the market site. Seek out a variety of local products and materials. Try to bring in vendors that offer products that do not overlap with one another so that each vendor has his or her own special niche.

The two main categories for vendors are established by the state: food and non-food. State permits for vendors are given out according to various requirements and restrictions for each category. It is helpful to organize them into these categories because different recruitment methods and permitting needs are required for selling different types of products. Non-food vendors including artists and artisans, are becoming more popular at farmers' markets. They attract customers looking to buy gifts for friends or to purchase organic gardening products. Additionally, farmers' markets do not need to be simply focused on selling food or non-food products to customers. Education and information tables provide the community with a service that will attract customers for different reasons. At The Food Project, we enhance sales for all vendors by inviting nutritionists and chefs to share nutritional information or to demonstrate healthy cooking methods.

Only ten more minutes until showtime!

Food Vendors

Attracting food vendors to a farmers' market holds incredible opportunities and several challenges for a market manager. At The Food Project, we have actively recruited farmers and gardeners to sell at our market each year. Several neighborhood gardeners join us weekly with their vegetables and culturally specific crops. A fruit grower from the suburbs of Boston has sold at our market since it began and he has developed meaningful relationships with the regular customers at the market. Periodically, we have also housed other vegetable growers.

Even with these successes, we also find challenges in recruiting for our market. One of the main issues at hand is the fact that our market is a subsidized market. A significant aspect of The Food Project's mission is to increase access to healthy, local food for inner city residents. We set our prices according to what people would pay for conventional produce in the supermarket. Doing this, The Food Project attracts many customers but also reduces the potential revenue of the market. In trying to recruit other organic or sustainable growers, however, prices greatly affect a farmer's interest and ability to join us at our city market. Many cannot afford to make the trip for such a small financial return and choose to go to higher end markets throughout the Boston area where they can guarantee high prices for their product. In addition, The Food Project grows a wide variety of crops very well. It is difficult to support more than one vegetable grower on the numbers we presently have coming to the market. Although many things have changed in Roxbury and especially in the Dudley Street Neighborhood, we also find that some potential vendors are concerned about the safety of themselves and their vehicles in what has historically been a high crime community.

A main key to increasing sales is to make the market look like an exciting place to shop. Over the years, The Food Project market has expanded with more tables and tents and a diversity of products to fill out and diversify the market. In 1998, we began selling transplants, non-toxic gardening products, compost, compost bins, bug books, and other gardening resources to the gardening community. We also piloted a fresh product booth, where The Food Project chef sold fresh salads, pastas, soups, and baked products with taste tests to customers. Adding this diversity increased sales at the market and encouraged people to return the next week for a different and interesting value-added product. The Food Project often buys products from local bakeries and other food businesses (i.e. honey) and sells them at a slight markup. This adds a niche to our stand and may encourage an increase in produce sales. We have also sold eggs, live broiler chickens, and other products consistent with the cuisine of the neighborhood.

Colorful produce and eye-catching signs add to the festive atmosphere of the farmers' market.

Consult your supervisor to discuss how much energy to put into vendor recruitment. The growers will supply you with the farm plan for the season and your recruitment can be focused on vendors that offer a niche product, something that The Food Project is not producing that season. Other vendors may share the vision of growing food for low-income families and may see participation in this market as a fulfillment of their business' mission.

Here are the steps to follow when recruiting food vendors:

- Refer to the Filemaker Pro Database. Search for "FM Vendor" in the ORGANIZATION Database in the Function field to retrieve all records. Check "Vendor" and "Ag Resource" functions in the ORGANIZATION Database and "FMarket Vendor" in the PEOPLE Database. Omit, if necessary, any records that do not apply. Print labels.

- Identify other potential vendors. Call the Department of Food and Agriculture or Federation of Massachusetts Farmers' Markets for any recommendations on growers who are looking for markets. The DFA publishes a directory of all the growers in Massachusetts and their products. Add any new contacts to the database.

- Revise vendor letter, vendor guidelines, and application to sell at the market (see Attachment 15: Sample Vendor Letter and Attachment 16: Vendor Application and Guidelines). Revise these guidelines as necessary, based on team meetings. Currently, we do not require our growers to be certified organic, though we would prefer a vendor with sustainable practices. This is an especially important clause for fruit as finding organic apples in our area may be challenging and they may be out of the price range for our customers. Be sure to include in the letter to vendors a list of things we can provide them with such as:

Attachment 15
Attachment 16

    - Tables. We have plenty of tables in the shed and should always have enough for other vendors.

    - Tents. Depending on the time of the season, we may have an extra tent or a vendor could squeeze under one of our tents. If the vendor does not have perishable items, they may be able to do without a tent. If this is the case, let them know they should be prepared for the heat of the sun.

    - Electricity. There is an electric source at the common. It is lo cated in the back of the common in the structure. It is in the center pole, up about seven feet. We also have several extension cords that we can supply to a vendor.

    - Stove. We have a portable "hurry hot" two-burner stove that we can make available to vendors who may need it.

    - Help. We can always offer a young person to lend a hand at the vendor table.

- Send letter, application and guidelines to all potential vendors in database. The letter should be mailed in mid-April and application should be returned by the beginning of May. Follow up with phone calls, encouraging visits to the site and to the neighborhood.

- Once the application and vendor fee is received, a follow-up letter should be mailed to confirm receipt. You may want to highlight key markets for certain events or other information you think is necessary.

- Call confirmed vendors in late May to confirm their attendance on the first market day (fruit vendors usually start later in the season). Ask the growers to predict the harvest date, so that you can inform the vendors.

- Thank you letters should be sent at the end of the market season.

Non-Food Vendors

We have found that vendors who are not selling food items are easier to attract to our market. Non-food vendors can fill a niche at the market, bringing their own homemade ideas and products. Non-food vendors need to acquire vendor permits, but are not subject to the handling and preparation standards that food vendors must comply with. Vendors who provide education or resources for the community are easier to recruit than growers or other producers. Most organizations would be excited to send a representative to the market to let the community know what they can offer. We also find that these kind of vendors complement our stand, in that they provide health and nutrition services, medical screening, entertainment, youth-centered community development initiatives, and so on.

A popular attraction for customers is the chef demonstrations, coordinated both by The Food Project and the University of Massachusetts Extension Program. The Food Project has prepared market vegetables in a variety of ways for tasting at the market. The University of Massachusetts has a farmers' market nutrition education program that provides chef demonstrations to farmers' markets and in the past we have had a chef on staff to do the demonstrations. Chefs use the vegetables grown by The Food Project and attract people through smells, sights, and ultimately taste. While people are tasting, they have the option to deepen their knowledge of nutrition and local food. Community booths are also a beneficial addition to a farmers' market. Be creative with your ideas and ask customers and neighbors what they would like to see offered at the market.

Here are the steps to follow when recruiting non-food vendors:

- Go through the database and search by FM for all Farmers' Market contacts. When choosing people to target, look for organizations providing a service for the community. Contact health centers, restaurants, food nutrition programs, culinary arts programs,

Department of Food and Agriculture, and university programs to interest them in on-site education at the market.

- Depending on your search, you may need to craft two separate letters; a general one to community groups who may want to do some outreach at the market and another letter to potential chef demonstrators (see Attachment 17: General Outreach Letter).

Attachment 17

- Update the application for vending in addition to the letter and send it to organizations and individuals. Provide vendors with an application deadline of May 1st, so that you have a confirmed calendar of events.

- Prepare a calendar to be aware of who is coming to the market each week.

Neighbor Gardeners

Today was the big day. At long last, Honorio was going to sell his produce at The Food Project's Farmers' Market. Honorio and the Urban Education Team of The Food Project had been waiting for this moment for almost a year. The process began when The Food Project received a grant from the EPA for an Urban Education and Outreach Program. In the program, several interns from The Food Project began working with local gardeners on land remediation, safe gardening practices, and creating educational resources for gardeners in Dudley St. Neighborhood.

Sylvia, a neighbor gardener selling her produce next to The Food Project's tables.

Honorio was the first gardener to reduce soil lead levels with us. He had a beautiful large garden with corn, squash, and beans that he used to feed his family and neighbors. It was upsetting, however, that Honorio was growing in lead contaminated soil, and we began remediating his land immediately. The process was long but successful. Honorio postponed his planting schedule while we added six inches of compost to his soil. Six inches would provide enough of a buffer so the plants wouldn't take in the lead through their roots. By the beginning of the summer, Honorio's land was safe to grow on. Even after this, our relationship with Honorio continued. Once a week, the Urban Education Team worked with Honorio to help him transition from conventional gardening to organic gardening. We identified bugs with him, and discussed organic methods to control the insect population. As we practiced organic methods with Hono-

rio, he taught us gardening tricks. For example, I learned that squash, beans, and corn are the three sisters and they work together in the garden. The corn provides a stalk on which the beans grow and the beans provide shade for the squash. By the end of the summer, Honorio's garden was completely organic and free of lead and we welcomed Honorio to sell at the market.

Honorio's produce was lead free and ready to be sold. That morning we washed vegetables at Honorio's home. We placed his table right next to ours at the market. As the market progressed, I became aware of what had been accomplished and what could be accomplished in the future as a community. We had helped Honorio remediate his land so it was not only safe for his family to eat but he could sell it at our market and profit from it. This illustrated the potential of forming a sustainable agricultural system where food was grown, sold, and eaten in the same community. Now at the market, his neighbors stop by and express their surprise that Honorio is there, selling his own vegetables, and discussing his practices in his native language of Cape Verdean Creole.

-Carla Campbell, Pollution Prevention Intern, 2000

When people have a connection to the land--they have hope in their lives.
John Steinbeck

A wonderful tradition at our market is the participation of neighborhood gardeners who sell their vegetables beside us at the market each year. Sylvia, a neighborhood resident and long-time gardener, has sold calaloo, collard greens, winter squash, tomatoes, and potted plants with us since the beginning of the market. During the 2000 season, we had another neighborhood gardener, Honorio, sell his vegetables at the market, thanks to an EPA funded grant to help eliminate lead in backyard urban gardens. Both gardeners have expressed the benefit they have experienced from participating in the market in the development of small businesses for themselves. We find that community members appreciate the presence of these two gardeners – since they are neighbors and friends. In addition, both gardeners offer vegetables that are specific to their cultural cooking.

It is important to define, with the rest of the urban team, who will be responsible for managing these relationships. The two logical choices would be the market manager or the urban education and outreach coordinator. You may find it easiest to be the direct contact with the gardeners since you will need to coordinate tables, tents, baskets, signs, and chairs. However, since the urban education and outreach coordinator works and communicates with the gardeners in the off-season, it may be helpful to stay in conversation and make sure that communication is not overlapping too much. Sometimes contact with the gardeners means stopping by their homes or jobs since it may be difficult to get in touch with them.

Gardeners have the option of selling their produce on either Tuesday or Thursday, and it may be helpful to you to encourage them to commit to one of the days. Work with them on a firm commitment around schedule, but know that things may change based on what is happening within their gardens. Be sure they also understand the coupon program and the use of the scales. The more welcomed and assisted they feel at the market, the more likely they will return week after week to sell their goods.

Refer to the section on setting up the market to determine location of neighbor gardener tables. We have found that over the years of piloting different table and tent arrangements, it works best to tack on extra tables at the end of our stand for the gardeners. Signs also help distinguish their small businesses apart from the rest. We want customers to be able to locate them, talk with them, and purchase their produce. Gardeners enjoy being on the outskirts of The Food Project stand, as opposed to separate and under their own tent, because they enjoy conversations and shared activity with the young people.

Special Community Outreach Projects

Community Outreach around your farmers' market has multiple benefits. By reaching out to established community groups, churches, or neighborhood based service organizations, you use already developed networks and relationships to increase your customer base. In addition, you are able to make use of services and resources available to these different groups to widen the audience for your market. For The Food Project, outreach to the community is essential to our mission and vision. Bringing various groups to the market brings richness to the youth experience as they learn about people. The community too is enhanced by hosting a market that brings in the very people that live in and around it in a deep way. The market itself becomes a place where community is enhanced and deepened, where relationships are cultivated, where connections are made.

The Food Project's Elder Pick-Up Program

The Elder Pickup Program is a well-established tradition at The Food Project's Farmers' Market. Every summer, The Food Project provides transportation to and from four local elder homes to our market. Elders living in area senior housing centers often find it difficult or impossible to walk to the nearest market to buy fresh, local produce. Therefore, The Food Project shuttle service increases access to vegetables and fruits for a population often limited in their food choices, activity, and mobility. The connection that has developed and continues to develop with seniors has strengthened The Food Project's ability to serve a diversity of age groups and provide an intergenerational experience for seniors, youth, and the broader community. The program is partially made possible with the distribution of elder coupons by the USDA. The interest in the market may change according to the money allocated yearly for the program.

The van pulls up close to the curb and out pops Claudette with a plastic stepping stool. She throws a smile at me before opening up the side door of the van. Twelve beautiful, wise women glance at each other inside the van, all patiently waiting their turn to exit onto the street. Four Food Project youth wait to hook arms with each of these seniors, so that they can stroll through the market stand. Silver hair, the handle of their cloth bag nudged in the crease of their bent elbows, hooking arms with the nearest teenager. In the bunch, there is always the leader, the comedian who keeps everything light hearted and fun, appreciating every moment spent out in the fresh air, amongst brightly colored vegetables. "Ooh, I sure do love those, and the peas last week, do you have anymore?" One woman turns to the other, "The greens were bitter last week." She is referring to the spicy salad greens in our lettuce mix. She puckers and shakes her head. A youth helping her tally her slip shrugs and smiles. Another youth offers a slice of a yellow doll watermelon. "Just like when I grew up down south. My daddy used to grow these." Makes me wonder how many of our young people have a deep connection with their ancestry, their history, their parents' parents.

The elder pickup program can be a logistically complex program. An effective and consistent system of communication is crucial to coordinating this program. It is important to locate a dependable contact person at each participating residence from the very beginning. Sometimes the contact person is a part-time staff who coordinates events. These positions often have a high turnover rate and, therefore, require time and investment from the market manager to cultivate new relationships from year to year. In other cases, the organizer is actually one of the elders who volunteers time to organize events and trips for the residence. The interest in attending the market is high in all residences that we have worked with – the elders appreciate the fresh food, the event, and the exposure to the youth. However, in order to make the program run smoothly, it is important for the manager to spend time on these relationships for a smooth and effortless program.

Here are the steps needed to execute the program:

- Recruit interested elder homes. Call the Elderly Commission and request an updated listing of all elder homes in the city of Boston. It will help future recruitment if you add any names and numbers to the ORGANIZATION Database. As you input these organizations into the database, use the organizational code field to mark them at "elder pickup recruitment", so that you can search for them at another time.

The senior citizens who come to the farmers' market appreciate the quality of the produce they can purchase at reasonable prices.

- Review the elder homes that have participated in the past. Go into the ORGANIZATION Database and do a find for "elder pickup" in the notes field.

- Before doing outreach to the coordinator at these centers, review the summer time evaluation from the previous year. This information may help to determine how many and which groups you offer the service to, based on staffing and resources for the next season. A good structure is doing two different pickups on each market day. Ideally, those two homes would be close to each other and the site supervisor would only have to go out once to do both pickups. In order to pick up at two residences, you might need to set limits on the numbers that can come from each place. Ideally, do this pickup as close to 4:00 P.M. as possible, otherwise the elders feel it is not worth their time to come to the market if we have already sold out of vegetables that they wanted.

- In April, contact the coordinators by phone and ask them if they are interested in the pickup program. Once you have a handful of interested participants, set up some parameters to work within, to make an effective schedule. This will require that you know the start and end dates of the pickup program, the staffing needs for this time period, and how many participants from each home on the average. Remember, in the past we have usually run the elder pickup only for the duration of the summer program (about eight weeks).

Attachment 18

- Once you have done this, you should revise the driving directions to each home (see Attachment 18: Elder Information) and take a trial run in the car. This will help determine when the site supervisor needs to leave in order to do the pickup and drop off in a timely manner. As you estimate time and distance, consider the time it takes to get in and out of the van and the time allotted for shopping based on your schedule. Assume that the person doing the driving knows nothing about Roxbury or Dorchester. Keep them on main roads with minimal turns.

- Type up the schedule, keeping in mind that the ideal would be to have them shopping between 4 P.M. and 5 P.M. Talk with the contacts at the homes to determine their needs for the schedule. Choose one time and one date that will stick for the duration of the program; changing the protocol in the middle of the program is a recipe for disaster.

- A complete schedule is one that works for you, the contacts at the elder home, and the person doing the pickup. Before finalizing the logistics, review the potential plan with the program production team.

- Draft a letter to each elder home outlining the details of the program and the times and dates of their pickups. Tell them what to do if they need to cancel a pickup; give them numbers to reach you and provide a deadline by which they must contact you if they need to cancel. Create a poster (often modeled after our general business poster) for the elder home to display in a common area. Leave a space to write the pickup day and time. Include some Farmers' Market flyers, so that people can see what will available at the market.

Setting up vegetables on a hot summer day.

- Schedule in a weekly call to the coordinators to remind them of the pick-ups. This may seem excessive but over the years we have found that program success depends on that depth of interaction. If you are working with a youth intern for the season, phone calls to elder centers build skills of communication, detail scheduling, community involvement, and integration with the elders.

- Make yourself available to the seniors in the program when they arrive at the market. Periodically check in with contact coordinators about how the pickups are going for them.

- At the end of the season, write all elder coordinators a note thanking them for their participation.

School Partnership Program

Connecting with local schools is another effective outreach strategy for a farmers' market. For the last couple of years, The Food Project has been deepening some of its relationships with the schools in the areas surrounding the market. We have been partnering with urban third graders and their science teachers on our food lots to provide hands-on learning about plant life cycles, agriculture, and hunger. As a part of the program of understanding agriculture and food distribution, special days at the market for children and their parents have been organized. These festival-like days include youth centered activities, special foods, and coupons for young people to bring fresh vegetables to their homes.

Outreach includes giving flyers to school partnership children during their school day. Coupons and flyers motivate the kids to come and bring their parents. (See Attachment 19 for an example of a market flyer for schools.)

Attachment 19

Operations

- *Staffing at the Market*
- *Market Internship*
- *Produce Availability and Pricing*
- *The Register, Cash and Coupons*
- *Summer Crew Worker Training*
- *Setting Up and Taking Down the Stand*
- *Running the Market Smoothly*
- *Events at the Market*
- *Keeping Business Records*

Staffing the Market

"One of my great memories is when our crew was at the Farmers' Market. People were praising us about how fresh the food looked. I took great pride in telling them that we had just harvested the food this morning. This was a great memory because it helped me to realize that I can help people with their needs. That I am useful."

- Brittany-Iris Scott, Summer Program Participant, 1999

Staffing needs for farmers' markets differs greatly from market to market. Traditionally, farmers and their families and friends staff market stands in larger markets. In the case of The Food Project, the staff for the market is comprised of youth. As the Farmers' Market Manager, you are responsible for training staff, managing operations, and ensuring the future of the market through proper record keeping and administration. The success of the market depends on the time, energy, and effort you invest in staff, as they will make the implementation and administration of the market much easier, if trained and deployed efficiently.

Staffing can be a difficult business. It is important to have enough staff to accommodate customers and keep the market moving. Too few workers behind the table can impact sales, frustrate customers, and decrease energy in youth and volunteers. However, overstaffing can cause youth and adults to work inefficiently and may foster poor role modeling and inappropriate behavior. When determining the number of workers for the market, consider the following factors:

- Volume and diversity of food harvested on the farm
- Weather conditions
- Customer needs

- Availability of youth at certain times of the year
- Management style
- Youth preparedness for running the market
- Whether or not a youth intern will be hired
- Number of well-trained support staff (youth or adults) that you have to support the needs of the market

Anticipating staffing needs for the market will be easier as you become more familiar with the market operations and the youth programs at The Food Project. While planning the market season, request sufficient time at Program Production meetings to talk about market staffing, training for youth, and overall market operation protocols. For spring staffing, talk to the Academic Year Program (AYP) coordinator to arrange schedules for participating youth. As the summer approaches, the market crew each week will be determined by the overall program schedule. When fall arrives, it is important to inquire with the AYP or Serve and Grow coordinator to discuss market staffing for September and October. The amount of time spent training directly relates to the level of experience of the participating youth. Expect that training hours will increase during the summer season when youth are new to the market.

A semi-quiet moment at the market.

Spring Market Staffing

The Food Project hires 60 youth for a summer job-training program and 22 youth for a leadership development program which runs during the school year. The market is staffed in June by youth who have participated in the Academic Year Program and have run the market previously. This is an advantage for a newly hired Market Manager. The seasoned staffing will allow you to troubleshoot operations and coordinate delivery, sales, and record keeping. Although these youth have been trained into the logistics of the market, it is a perfect time to deepen their understanding of the enterprise, local food systems, and customer service. In the past, we have given significant responsibilities to the experienced youth in the spring – leading all aspects of the market from setting up the stand, displaying all vegetables, and weighing out all the produce.

She would hop off that bus, every Tuesday and Thursday, just in time for the market to start. Sneaking up behind the stand, she would put her hands over my eyes, and wait for me to guess who she was. It was never difficult for me to remember Vanusa's presence after that moment, as she would grab a bib, stuff a pen and sales pad in the front pocket, and head to the stand to intercept the customers speak-

ing in her native language. I was never really sure what was said, but I so trusted this young woman, and felt as though in her own language, she was serving the customer with the best that The Food Project had to offer. She was a quick learner. She negotiated price changes in Portuguese. She was honest with the customers and knew her product. She was always ready to learn more, recognize a mistake, and correct it for the next sale. We would push each other to understand the deeper meaning of what we were doing together each market day. Vanusa felt confident because of her continued desire to learn more. I felt privileged to have a young woman working for me, who was interested in more than just selling food. She perfected her job, a tribute to ongoing training and an internal desire to succeed.

The AYP Coordinator will help you identify youth able to work on Tuesdays and Thursdays after school during the school year. Raise any staffing concerns or logistics at the Program Production meetings in April and then again in August for the fall markets. Up to three youth are needed for the spring markets, when there is less variety and volume. In the fall, on the other hand, when the farms are booming, it is important to have at least five youth staffing the market. You will get a sense of staffing numbers as time progresses. Do not hesitate to let young people go early, if it seems feasible. It will be important that you encourage young people to contact home when they are leaving early or staying late, as parents might be concerned for their child's schedule. This is especially important in the fall, when darkness comes earlier in the day.

Summer Program Staffing

"The farmers' market is a relaxation from farm work. I personally like it, because you have a chance to see the produce that you weeded last week go into the bags of middle-aged men or women. We got a round of applause for doing such a great job."

- Maria Vieira, Summer Program Participant, 1992

Staffing needs change as the summer arrives. At The Food Project, the market is considered one of the highlights for youth in the summer program. The markets require high coordination of Program and Production Staff. This coordination of schedules will be worked out during the early spring meetings with the Program Production Team.

Youth are introduced to the farmers' market for the first time as they participate as crew members in the Summer Program. The farmers' market orients young people to a real business, helping to build basic skills in customer service, math, and public speaking, as well as shed light on The Food Project's mission in working to build a local food system. The Market Manager is responsible for facilitating this experience, offering a strong presence throughout the summer, and managing skill building. Be prepared to shift your schedule so that you can accommodate the added hours necessary for both training and operations.

Each crew is composed of eight crew members, an assistant crew leader, and a crew leader. During the Summer Program, each crew runs two markets, the first market on Thursday and the second market on Tuesday. Ten youth is a large number to run a market stand. Developing systems for efficiency and identifying tasks that can be done in addition to running the stand should be one of your primary concerns during the summer markets. The training sessions for running the market are outlined in *French Fries and the Food System*, a Food Project curriculum guide, which should give you a sense of how the training works for each crew.

In general, training for crews is as follows:

- A crew arrives on Thursday morning prepared to participate in the first training workshop. This takes up to two hours. The training includes the basics of: completing a sales slip, vegetable identification and pricing, weighing and bagging produce, and working with a sales partner, among other tasks.

- On the following Tuesday, members of the same crew will visit another farmers' market in the area, speaking with other vendors who sell at farmers' markets.

- When the crew has worked both the Thursday and Tuesday markets, the Manager leads the crew in a one hour Market Analysis, moving the youth through the bookkeeping and accounting tasks for market administration.

I am convinced that my life belongs to the whole community, and as long as I live, it is my privilege to do for it whatever I can, for the harder I work, the more I live.

Melnea Cass

The Market Manager's responsibility includes organizing, giving feedback, and being patient. It is important to be cognizant always of the crews, making sure that they are being courteous to customers, calculating sales correctly, and working well together. When the youth are running their second market, raise your expectations for their performance. Work with them to improve their customer service skills, encourage them to anticipate the customers' needs and take greater initiative with restocking items on the stand.

The Market Manager is also responsible for noting Standards and Straight Talk issues (see *Growing Together* for a description of this technique). The young people are still working under the standards agreement with The Food Project and should be held accountable for any behavior issues.

Staffing Between Summer and Academic Year Program

"Under the youth leaders' instructions, we picked tomatoes and squash and basil and six other vegetables, washed them, and packaged them for the farmers' market. We will never forget the days we spent on the farm, as I'm sure no student who comes here will. The land, the people, and most importantly, the chance to produce, prepare and sell – all joined to create an invaluable experience for us."

- Vocational Teacher, Concord Assabet Adolescent Services, Inc.

There will be times when you must locate volunteers, Food Project alumni, or interns to run the markets. Alumni youth are usually the easiest to integrate into the market since they have run it many times and are clear about the protocols. Many of the alumni are excited to come back to reconnect with the neighborhood and spend time at the market again. When switching staff throughout the season, it is important to your customers to keep the style, set-up, and rules of the market consistent. Product and service quality and consistency are critical features to the success of any business. With this said, however, volunteers, especially those who have a long relationship with The Food Project, can greatly benefit from working at the market. In doing so, they see the end goal of the work they have done in the fields with the youth and staff. They meet the faces of the people that benefit from local, organic produce. They are able to interact with the diverse clientele that frequents the market. If you decide to use volunteers, they should be trained, in the same way as the youth, in proper protocols and procedures for operation. This necessitates that the manager meets with the volunteers before the market for training. In order to schedule staffing for the in-between times, talk with other staff 3-4 weeks prior to the dates and determine the best options for the market.

Market Internship

My name is Janique. I am 15, from Peabody. I have been with The Food Project since the summer of 2000. I have been through the sum-

mer program, the DIRT crew which stands for Dynamic, Intelligent, Responsible Teenagers, and now I am part of the internship program. I have stayed with The Food Project because, at each level of The Food Project, I learn something new about myself and other people. This summer, I am the Farmers' Market intern. During the winter of 2000, I proposed a Farmers' Market internship to The Food Project staff. I did this because I am really interested in the community and I felt that the Market was a central place to reach out to the community. I loved the Farmers' Market from my first summer and I wanted to feel like I was doing more. It is important for me know that the organic vegetables that we grow are being sold for inexpensive prices to people who really need them. As the Farmers' Market intern, I train crew workers into running the market, I analyze the market with the crew workers, and I go to Stop & Shop weekly to find out prices for vegetables and fruits. I also work with the money that comes into the market, calculating how much money was made and how many customers came to the market that day. At the beginning of the summer, I set goals for myself. I wanted the market to run well so that we would make money and reach the goal we had set for ourselves. My second goal was to make sure that people understand where their food is coming from and that it is healthy, organic and inexpensive. I wanted people to know that you don't have to travel a really long way to get fresh food. Doing the market internship this summer, I have grown tremendously. The more I do with The Food Project, the more I will grow as a person.

– Janique Parrott, Age 15

Janique Parrot, The Food Project Farmers' Market Intern.

Integrating diverse people brings richness and life to any market. Food connects people across ethnic, racial, religious, age, and class boundaries. Vegetables, fruits, and local products especially have the potential to bridge gaps between different people because the buyers can be connected to both the land and the growers who produced the food that they purchase. Having the farmers and producers selling the fruits of their own labor and care in the fields brings a very particular experience to a farmers' market. At The Food Project, youth are the sales people and the farmers. They spend their summers working out on the fields, harvesting and washing vegetables, and bringing the vegetables to their points of distribution – the shelters, the Community Supported Agriculture Program, restaurants, the kitchen enterprise, and the market.

The mission of The Food Project is "to create a thoughtful and productive community of youth and adults from diverse backgrounds who work together to build a sustainable food system." Through working in the fields, attending workshops, running the market, and assisting at shelters throughout the Boston area, the youth at The Food Project are introduced into this productive community. After the initial summer, the alumni youth are invited to stay connected with the work they have participated in for two months and to continue to contribute in the area that interests them. There are several opportunities for youth to stay involved in building a sustainable food system. One major way that young people can stay involved is through the Alumni Internship Program that offers advanced positions within the organization. In these alumni positions, the youth are challenged to participate on a deeper level with the work of The Food Project. They become part of the planning process for workshops and activities, take a leadership role in carrying out projects, improve their skills of communication and presentation, and cultivate a deeper understanding of the issues around local food systems.

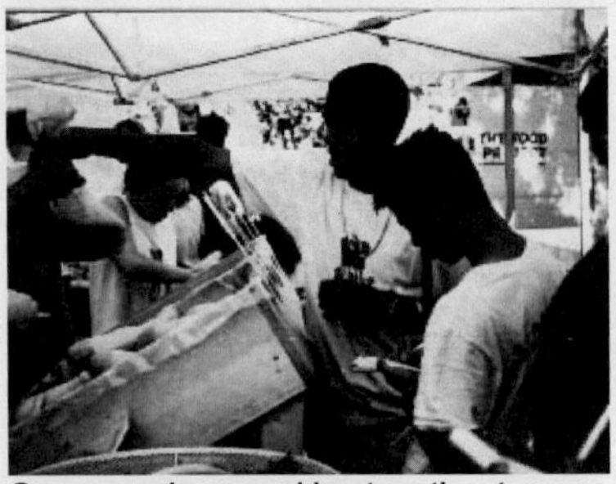

Crew members working together to complete a sale.

In 2001, the Farmers' Market internship was created in response to a written request from one of The Food Project alumni. The market had been a powerful part of her summer experience and she wanted to be more fully integrated into the preparation, marketing, record keeping, and training for the market. Integrating a young person meaningfully into a position is not to be taken lightly. It is important to assess the youth's skills – strengths, challenges, and interests. At The Food Project, we want the youth to be able to shine in their position using their natural gifts and abilities. However, we also seek to challenge youth to take on responsibility and leadership that might not come naturally and to develop useful and meaningful work ethic and skills. As market manager, you should brainstorm market tasks that are important, interesting, educational, and manageable for a young person. Some jobs are more defined than others and allow the intern to take a great deal of responsibility for them. When tasks are completed, the intern can report back to the manager. Other tasks are more involved and nebulous and will need to be done in conjunction and collaboration with the manager.

Below are some of the tasks that were itemized for The Food Project market intern during the 2001 season. Use this list only as a guide. The manager should determine the best arrangement for the specific youth and adjust or change the tasks accordingly.

- **Price Manager.** As the season progresses, vegetables and fruits come and go at the farmers' market. In addition, the "running rate" for vegetables also changes according to the week or month. At The Food Project market, we sell our organic vegetables for a comparable price to the conventional vegetables sold in the supermarket. We do this to enable people regardless of income to benefit from access to local, organic food that is healthy and safe for themselves and the environment. A visit to the supermarket each week during the season is an effective way to set our own market prices. Dependent on whether visits to the supermarket are a possibility for a young person, the market intern can be put in charge of the pricing of the vegetables. The intern can type up a price list and change it weekly for the trainings and for the market itself.

- **Local Advertising**. The intern can be put in charge of sending out PSA's to local newspapers every other week. There are several different valuable skills to be developed through this job. First of all, the intern makes a contact at each of the local papers and begins to understand media scheduling and restrictions. In addition, the intern becomes familiar with the office's computers, fax machine, and phone system. Through altering or creating the announcements, the market youth is challenged to write concisely and clearly to stay within the word limit. Finally, the youth also is exposed to the importance and activity of advertising for the market throughout the season. Note: The form for the PSA is Attachment 11. **Attachment 11**

- **Market Training Assistant**. Each week, a new crew is trained into the market. The training includes weighing and bagging, customer service, vegetable identification, and money exchange. During the training, the youth also learn a bit about the history of the Dudley Town Common Market and the philosophy behind the market. The intern and the manager run this training together. While planning the training, identify parts that the intern can explain and run. The training itself is diverse and active and holds a lot of potential for the development of public speaking and leadership skills.

- **Produce Display Point Person**. A significant aspect of a farmers' market is display. At The Food Project, we strive to have our tables clean, attractive, colorful, and organized. We also keep our display consistent so that our regular customers know where to look for carrots, broccoli, or garlic each week.

Setting up the market is a very active and bustling time. The market intern can be responsible for the display of the vegetables on the tables while the market manager oversees other parts of the set up process. The intern can direct the youth who are unloading the vegetable trucks and then oversee the placement of the produce on the table. Doing the job from week to week will keep the display consistent. The intern develops skills in directing other people, in creating an attractive and functional display, and in knowing all the produce that is being sold at the market.

There are many other tasks that can be taken on by the market intern. It is important that for some of the jobs, the manager be present. For example, everyone is interested to know how much cash and how many coupons were brought in at each market. It is great training for the market intern to count and record the money and determine the customer tally for the day. However, in all cases involving money, the young person should have the protection and accountability of a staff member present during the accounting process. It is important to develop and hold trust in one another during all stages of the market. At the same time, it is necessary to always avoid circumstances that put young people in situations that are difficult to handle.

The market internship for 2001 was a great success. When two capable and interested people are working together on a project, they are able to accomplish more and do a better job than if either of them were working on the task alone. The intern becomes a role model for summer program youth at the market while deepening his or her personal and work skills. With a summer of working in most aspects of market management, the intern becomes deeply invested in the work that is being done. During this process, guidance and support from the manager is critical. The one-on-one working relationship between the Manager and Intern promotes learning and growth on all sides.

Produce Availability and Pricing

Having a consistent product for a farmers' market and fair pricing are essential components of managing the market. To begin with, the Market Manager should know what crops are being grown and during what time period they will be available. The Manager should serve as a resource for the customers and the young people at the market. Customers will seek you out for this information from week to week. The growers who deliver produce from Roxbury and Lincoln and will be able to supply you with specifics around harvest dates for popular items.

Spend as much time as you can on the farm, in the gardens, and talking with growers to build your knowledge about agriculture and food. This information will help you talk with customers about the food at the market. The agriculture manuals serve as a resources for any agricultural questions asked by customers or youth. Recipes and nutrition materials serve as ways to initiate conversations with customers about local food and health. If you don't know the answer to a question, find the answer, and let that customer know the next week. Your connection with the growers through the season will serve as a base for communicating information to customers about what they can purchase in the future.

Before the market season, develop an efficient and convenient process for communicating with growers about product availability. Some suggestions to set up a system of determining the products for each week for the market:

Peppers harvested on the farm in Lincoln for the CSA and the Farmers' Market.

- Meet with both farmers before the market season (April or May) to determine when you will communicate about produce. This should happen twice a week, either Monday or Tuesday morning for the Tuesday market and then again on Wednesday for the Thursday market.

- First, talk with the urban grower because their crops are grown specifically for the market. The Lincoln crop plan is planned to supplement the produce grown in the city (Lincoln produce is grown specifically for shelters & a suburban Community Supported Agriculture Program). Find out what the grower has available and how much (for example, three crates broccoli, five bunches of basil, and so on). Write this down. Then call the Lincoln grower to communicate what you will receive from Roxbury and what can be supplemented from the Lincoln farm. Since certain vegetables are only grown in Lincoln, it is important to connect with the Lincoln grower regularly.

- In your meetings with the growers, discuss the general quantities that the market can handle during different parts of the season. Review the prior year's records to see how the market changes over time. During the season, pay attention to each market in order to determine if we are bringing in too much of a certain vegetable and are consistently selling out of another. We may only have a limited quantity of certain vegetables to bring in (for example, corn) whereas there may be plenty of another that can be brought in if the customers are demanding it (for example, collards).

- Create a list of available vegetables to be faxed to local health centers and other community organizations for posting. Many centers have inter-office email and will communicate with each other. Make a contact in the health center who will do this for you. This method may work well at 1010 Massachusetts Avenue, where you'll find many city of Boston departments, such as Inspection Services, Parks and Services, and Human Services.

- After determining what is available for the week, you should go to the local supermarket (for example, Super Stop and Shop at South Bay Plaza) and note the prices of all items you are selling at the market that week. You may be able to price some organic produce, but most of your price checks will be from the conventional produce. Usually, we keep our prices at or slightly higher than the supermarket price for the conventional items (not organic). In the past couple of years, we have tried to increase our prices slightly and the customers have not rejected this. In general, these customers cannot afford organic prices, but will pay a little more than the supermarket because they know they are getting fresher produce.

- Round the prices to the nearest quarter dollar to make things easier for youth running the market (for example, if something is \$1.39 a pound, you might round the price to \$1.50 or \$1.25).

- When you begin pricing items, check with the growers; you may want to discuss the prices for the upcoming year based on the experiences of the previous year. Get the growers' feedback early and then continue discussing it with them throughout the season. Get their views on the way each item is being sold (by the bunch, pound, bag, or head). Standards are already set up for this that the growers adhere to when they harvest. Remember a cardinal rule: if no one complains about prices, then it means they are too low. You know your prices are fair if a few people are complaining. If lots of people are expressing frustration, then prices are too high.

The Register, Cash and Coupons

"My goal for the summer was to learn about the Farmers' Market and I do feel like I have achieved this. I learned I can make a difference and I am patient and tolerant, adjustable and giving. I did have challenges too. I guess rude people could scare you away from learning and having fun. I learned that I am not a great salesman, but that I work very well at the register."

- Joseph Mazzarelli, Summer Program Participant, 1997

At an active and bustling market, the coupon and cash exchange can be a tricky activity. Since farmers' markets can get very busy, there is a lot of activity at the cash box. It is easy to be careless while trying to give customers the quickest possible service. Yet, it is important for records and market sustainability to make as few mistakes as possible with money exchange.

Counting cash for the day's market.

At The Food Project, we have always had one cashbox at the market, which is filled with cash and change at the start of each market. Receipt pads are kept at hand for the youth to add orders and keep track of their sales at the market. Near the cash box, there is a customer tally to keep track of the numbers of people that purchase vegetables from us each market.

Over time, we have entertained the idea of various methods of cash collection. For part of one season last year, we obtained a cash register, which assisted in adding up the orders and keeping track of the number of transactions at each market. The cash register allowed us to total thc sales for each market and create a day's receipt to match the record totals to the money in the cash box. This method assisted in catching any mistakes or theft. The cash register also gave an accurate count of the total customers for each market. Of course, in considering a cash register as an option, you need to guarantee an electrical source at the market site. In addition, the manager also needs to be ready to do maintenance on the cash registers if they are to break, run out of paper, or jam.

Adding up customer amounts quickly can be a challenging task for youth. In the past, we often used the market as a way to train the young people in some basic math skills. Overtime, we have come to prioritize expediency for our customers. There are other parts of the market training and analysis that can focus on the development of quick and accurate math skills. In order to facilitate this adding process during the market, it is helpful to have basic adding machines so that totals can be figured quickly and accurately. In addition, if each youth has a calculator at hand, totals do not need to be computed at the cash register, which will allow for a smooth flow past the cash register. The process for cash exchange is as follows:

- The youth who are acting as salespeople take orders, calculate the overall amount of the purchase, and ask for this amount from the customers.
- Salespeople then bring the cash to the cash register, telling the cashier the amount spent and the amount that is being given to them.
- The cashier enters the total into the cash register to make a record of the sale, and makes change for the salesperson.
- The salesperson takes the change and returns it to the customer.

Beets! Only a dollar per bunch!

Other protocols for using the cash register include:

Each market day, start with $100.00 in the cash register with this distribution:

- 13 — $5 bills
- 18 — $1 bills
- 1 – Roll of quarters ($10)
- 1 – Roll of dimes ($5)
- 1 – Roll of nickels ($2)

By the time the market ends, it is too late to deposit the cash at the bank. It is possible for the market manager to obtain a night deposit key and drop off the cash at the end of the market. There are advantages to this option: the money is out of the manager's hands and there is no danger of losing it. In addition, since the manager's position at The Food Project is a part-time position, depositing the money on the same day saves coming in on a day when there might not be other obligations to attend to. However, choosing this option means counting the cash and recording it on the deposit slips the night of the market. By the time the market closes, the site is cleaned up, and the vegetables carted away to the shelters, the youth, the grower, and the manager have all had a long day. It is important to return to the office or another safe place to count and record the money. All of these steps simply add time to an already long day.

Another option is for the manager to return to the office and lock the cash in the urban office. Storing the money in the office must be done carefully and safely. Set up a system for this with the office manager right away. Keeping the money in the office until the end of the week allows the cash to be counted and recorded during the youth market analysis. At the end of each week, the youth qualitatively and quantitatively analyze the markets that they have run. The cash is one important part of this and the youth benefit from having it there for accounting purposes.

The office manager will provide the manager with deposit slips for the market currency. Keep this book of deposit slips in a convenient place. Set up a consistent system for bank deposits. Sometimes it is easiest for the manager to bring the deposits after the market analysis. It is also possible that the office manager could deliver deposits during other trips to the post office and the bank but the logistics of this must be worked out early on.

Accepting Coupons

The market coupon program of the USDA is a valuable resource to low-income children, women, men, and the elderly. A farm must register with the Department of Food and Agriculture (DFA) to be able to accept WIC and Farmers' Market coupons. In the beginning of the spring, register our farms so that we can accept coupons at our market. WIC and state dollars are contingent on funding each year. The Food Project has accepted food stamps in the past, but as of 1998, we only participate in the WIC Program because of the logistics of food stamp redemption.

To accept coupons follow this procedure:

- In April, contact the Farmers' Market Coordinator at DFA to check on the guidelines for participation for the coming market season. The Federation of Massachusetts Farmers' Markets (FMFM) is also a great source for information on coupon guidelines and schedules for payment during the season. You can find contact information for both groups in the database.

- By mid-April, one of the organizations should have sent you a Farmer Certification Guideline Overview and an application to participate in the WIC Program. They will usually do this automatically based on our participation in the past. Call them if you have not received any notice. Typically, May 15th is the deadline for application that enables markets to accept coupons.

- When you apply for the program, you will be eligible to accept $2.50 WIC Coupons, $2.50 Farmers' Market Coupons, and Elder Coupons. The color of these coupons varies from year to year and sometimes the amount that these coupons are worth varies as well. The DFA will send you a letter outlining the details of the coupon program for that year.

- You will receive a packet from the DFA when the DFA has received and approved your participation in the coupon program. In the packet, you will receive small business cards indicating your participation in the program and two important numbers for redemption; the Market Certification Number and the Market PIN. Keep these cards and numbers on file for use when redeeming coupons.

Spraying the produce keeps it fresh and cool.

- At this point, you may also begin to receive correspondence from the Federation of Massachusetts Farmers' Markets. The Federation is contracted by DFA and handles coupon redemption. Redemption correspondence takes place with the Market Operations Coordinator at the Federation for Massachusetts Farmers' Markets. Request for Payment information and forms as well as a redemption schedule will be mailed in the acceptance packet from the DFA. This schedule will also indicate the last possible date to send coupons in for reimbursement. In the past, the last date for FMFM reimbursement has been early November.

- The FMFM encourages all markets to redeem coupons throughout the season instead of saving them all to the end.

Steps for Coupon Redemption:

- To redeem coupons, be sure you have done all the record keeping for each market, and separate the coupons into three piles by the type of coupon. If we set up a market at another location, then you will need to redeem coupons separately for each market. It is important to keep separate records for each market location for redemption.

- Once you have separated the coupons, count them and record these numbers on the 'Request for Payment' form. Fill in other information on this form including Certification Number and PIN number, which were given to you by DFA.

- After completing this form, send your stack of coupons to the Federation of Massachusetts Farmers' Markets. To reduce the chance that this packet will be lost in the mail, use UPS, who can track the package, or register the package with regular mail.

- A check will come to the Financial Administrator at The Food Project and will be deposited by him or her. You should request that a copy of each check is given to you for your files. The money for this check should be recorded on the sales database immediately.

Summer Crew Worker Training

"The farmers' market shows us many skills including working with people, managing yourself and others during chaotic times, and being patient."

- Yveline Alexis, Summer Program Participant, 1996

Training for a market crew in customer service, vegetable identification, weighing and cash exchange, among other things is an essential process for any farmers' market. The Food Project's mission is to work with youth in all aspects of the market and this goal necessitates thorough training for all participants. Over the years, we have set standards for how we expect our young people to behave at the Farmers' Market. See the section on "Staffing the Market" to review expectations we hold for The Food Project youth. The market is one of the only places where the Summer Program participants act as ambassadors for The Food Project. For this reason, we expect excellent work at the market from both staff and youth. Role modeling is essential for the success of the market process and any behavioral issues should be dealt with immediately.

At The Food Project, we expect that all youth are honest and trustworthy. We give young people a lot of responsibility at the market and assume that they will see this as an opportunity to be challenged, held responsible, and given leadership opportunities. It is important that the youth know the level of responsibility being given to them. Giving clear instructions and jobs is an effective way to encourage accountability and responsibility at the market.

Steps in preparing youth for the market:

- Determine the Summer Youth Program dates and verify the Roxbury Schedule with the Summer Program Coordinator by March. The Roxbury schedule should include a timeslot reserved for market preparation. This schedule should include ample time for training on all of the aspects of the market experience. When planning your training sessions, use the former year's schedules as a guide for planning and implementing your training program.

- Having informative, interesting, and interactive activities to train market workers is important. Read through the Farmers' Market training section in *French Fries and the Food System.* You will find a series of three lessons for the training of summer program youth. The first lesson in the curriculum focuses on training young people into the operations of the market. Always begin the lesson by describing the market and giving the context. It is important for the youth to know *why* we run the market, *when* we began it, *who* it serves, and *what* it offers to the community. Often the reasons and background for the market can come from the youth themselves, so try to make the exercise interactive. What might be the advantages of a market in this area? Why does The Food Project commit to working in this neighborhood? When the youth have an overview, it is time to move into the specifics of running the market. The youth are divided into learning stations – money exchange, weighing and bagging, vegetable identification, and customer situation cards. Each group has a leader who guides them through the important aspects of each station. The interactive curriculum brings the youth directly into real market situations and helps them to practice what they will be doing later in the day. This type of training requires help from others, specifically an experienced youth for each training group. In the past, advanced interns who have run the market previously have assisted in the market training. You may need to do some preparation in order to find advanced young people to help you. When you have isolated four youth to partner with you in the training, take some time to work with them to make the training flow and make it fun for all.

- At The Food Project, training does not just include direct preparation for the running of the market. The second lesson in *French Fries and the Food System* is the Market Visit. Each crew is taken to another farmers' market in the metro area. The visit exposes the youth and staff to different markets throughout Boston. Seeing other markets prompts brainstorming on and improving our own market – presentation, product, customer base, pricing, and various other market aspects. Since The Food Project is committed to the Roxbury community, it chooses vegetables, operation times, and prices accordingly. Because other markets have different customers, locations, and missions, they may set up, advertise, and run the market differently. It is helpful for us to compare our methods and see if we can adopt any of their successful strategies. (See Attachment 20 for a question sheet for market visits.) The Food Project has visited the Newton market for many years. It offers a wide diversity of products, the farmers are often very willing to talk with the young people, and the market manager is seasoned and organized and willing to give our group a tour. A two hour block is usually set aside for the market visit, including travel, tour, visit, and discussion.

Attachment 20

Who wouldn't love the trip to the Newton market? It has all the wonders of a field trip – mystery, discovery, newness, and treats. On steamy days, it is a half an hour to relax in the van, an opportunity to have a cool drink or snack. It holds the excitement of seeing a new place and the power of a myth, told by past crews of ice cream, fresh baked goods, and cold drinks. Beyond the novelty and the excitement, though, there is a beautiful old maple tree that spreads its shade over to the side of the grouping of 20 farmers' tents that make up the Newton market. It is in this space that TFP youth gather in the cool grass, with ice cream cups and cones, and talk about fresh vegetables, organic production, class and race issues, food security and justice.

Every week, Junique, the market intern, led the Market Visit preparation as we drove through the construction-laden streets of Roxbury and Brookline, over to Rte. 9 to lead us to Newton Heights. She handed out bright orange half-sheets of paper entitled "Things to Think About" to each of the crew members and together they would read down the questions. At this market, how are things displayed? Are the vegetables at the market organic? What is the range of prices for the vegetables on the stands? Where were the vegetables grown? What types of people come to this market (race, age, economic background)? How many of the stands accept WIC or Farmers' Market coupons? She would divide the crew into 5 teams of 2 and each group would roam freely but with purpose, all of them responsible for bringing back observations, answered questions, and new ideas for our market to the rest of the group.

A neighboring farm, Drumlin Farm, holds a farmers' market.

Crew D was psyched for the trip. Smack in the middle of the summer, Junique and I had had time to perfect the trip: we knew what to expect from the visit, we had gotten to know the growers, we knew where we'd

meet the most traffic, when we should leave Newton in order to get back to Roxbury in time. We also had developed effective guidelines for the crews, a mix that began with a formal tour of the market by the manager, but also allowed for some freedom for observing and asking questions. Janique and I had run the discussion several times and knew what issues we wanted to raise if they weren't brought up by the youth. Much of that discussion revolved around comparisons and contrasts – our "convenient, affordable, coupon accepting" market compared to the "other," the upscale Newton market. And, yet, with all this, we still were not prepared for Crew D and the Newton-born, Alex.

Alex sat in the front of the van and offered his services as a navigator. I pretended that I wasn't quite sure of the route. When we arrived, Alex was not to be contained. He kept running into people he knew, hugging his neighbors, old elementary school friends, and teachers. He would try to introduce his TFP crewmembers to the plethora of faces that joyously called his name, but Crew D was off and about, talking to vegetable, beef, goat, and mushroom farmers. Alex was a little late making it back to the maple tree shade where we had all gathered and were discussing the differences seen in pricing here versus our market. He was reeling from the excitement of being a star in his home.

"Why are the prices higher here?" Janique continued the discussion as Alex sat down.

Emma piped in. "Because we price them according to what our customers can pay. The people here are able to pay more." "Yes," Courtni continued. "And we only charge what our customers would see in the grocery store, even though our food is local and organic." "But I don't think their produce looked any better than ours," Josh said. "And most of the farmers weren't organic. I only saw one that was. But still, I think they are charging too much."

"Let's think for a minute why they are charging so much," I said.

"They are trying to make a living," Brendan offered. "Yeah," the crew leader, Kevin, began. "If we figured out all of our expenses, with staff, land, materials, time, and youth, we definitely do not make enough money in our cash box to cover it. We are able to offer vegetables at our prices because we have other sources of money, isn't that right? And because our livelihoods don't depend on what we bring in on Tuesdays and Thursdays."

"And what does that say about sustainability – in terms of economics?"

"Well," Kevin continued. "It is important to be economically sound as well as environmentally sound." "Maybe we could add more products to the market," Brendan suggested. "And make more money." It was agreed all around that ice cream should be a priority. "Even so, though," Alexia said, as she finished her cookie, "I don't think we should raise prices because it is part of The Food Project to make this food accessible to the neighborhood. And I only saw one stand that took the coupons. We should always take the coupons." Everyone nodded. It was a hard mission to compute – being sustainable and fulfilling our commitment to the neighborhood.

Alex was getting restless. "One thing I noticed," he said quietly (and this wasn't a usual way for him to communicate). "Is that I only saw white people here. Newton is mostly white. At our market, we have people of different races and cultures. The people who come to our market speak different languages." He paused. "I know it sounds simple. But it is really hitting me today." One of Alex's teachers called out his name and gave him a huge wave. Alex smiled. He repeated, "It is really hitting me today."

The "other" – the Newton market – was humanized in a new way for all of us that day. With Alex's epiphany about the seemingly vast differences between what he was doing day in and day out with The Food Project on the land, at shelters, at our market, compared to where he was from, carried us to a new depth of complexity on that sunny Tuesday. Here we had touched upon economics, culture, race, access to food, and class dynamics and we had each other to bridge the complicated gap that we were seeing between two worlds. We stepped out of the shadiness of the maple tree with a lingering sense of how our very lives and backgrounds play into our (and The Food Project's) positioning in a food system, in a culture, in a society.

- The final lesson for the market is the analysis. Every week, each crew has the opportunity to qualitatively and quantitatively analyze the markets that they ran. The analysis is also very interactive, having youth count the cash and coupons, tally up customer numbers, and brainstorm on what worked and what did not throughout the course of the market.

- In May, read through the lesson plans and revise all necessary worksheets and tools used for the market training. By June, you should have contacted farmers' market managers for the field trips.

Setting Up and Taking Down the Stand

Unloading greens for a market.

Time and efficiency is crucial in set up and take down of the market. At The Food Project, both of these are time sensitive and involve managing several young people as well as several pieces of equipment. Therefore, the more organized you are in the management of the young people during the set up and the take down, the better it will go. Several suggestions for how to organize people during these two important times follow. Read these over, try them out, and make any changes that you feel would be easier for you and would help move things along quickly. Each market manager, based on their personality and skill set, will add different dimensions to the Farmers' Market and therefore it is crucial that you find the best ways to make the market operations work for you. Also, it is important to communicate exactly and directly with the young people about what you want them to do. It may be effective to give the youth partners to work with and make them accountable to one another in each of the tasks. If it is feasible, create short but detailed "To Do" checklists for each pair so that everyone is clear as to what they are doing.

Set Up

As part of the training, the manager should divide the youth into set up and take down teams. For set up, there should be a group of four in charge of setting up the tents. A group of three should put up the tables, sandwich board and banner. They should hang the scales, chalkboards, and bags while the tent group moves to set up the vegetables on the tables. Set up proceeds as follows:

1. Load all of the Farmers' Market equipment into the market van. The following is a list of or the items you will need at each market:
   - Tents (2-4 depending on the season)
   - Tables (4-6 depending on the season)
   - Fold out chairs for gardeners
   - Bin with paper and plastic bags (full)
   - Sandwich boards
   - Bin with hanging scales (4)

- All of the bushel and half bushel baskets, some smaller ones
- Bins with materials (pens, markers, receipt pads, rubber bands, tools)
- Metal tubs for displaying veggies with water
- Bin with vegetable signs
- Water cooler filled with drinking water
- Chalk boards and chalk
- Signs to prop against tents
- Cash registers and calculators
- Broom and dust pan

2. At the market site, unload all market equipment and supplies. Unload the tents first and have the tent team set them up while everyone else is unloading the remaining equipment. One person can stand at the back door and hand items to people and another person can stand at the side door. No more than two or three people should unload, as it gets crowded at the back door of the van.

3. Have each member of the tent group grab a tent leg and first pull the legs out as far as they will go. Then, they should step on the tent foot or bolt on the bottom of the leg and push the leg up from the bottom. Each leg will snap into place once on the bottom and then again on the top. Make sure that the pins are in the same holes for each leg. It is important that this happens first to provide shade for the vegetables.

Once a market is set up, customers are eager to shop.

4. Note: The E-Z up tents used for the markets are sturdy and functional but they are light. When the weather is windy or stormy, the tents are easily blown around by the wind. Depending on the severity of the conditions, different strategies for anchoring the tents can be used. One way to prevent the tents from shifting around on a light wind day is to tie the tents together with bungi cords or rope. With all the legs of the tents connected to each other, the wind cannot pull up the entire structure. However, on blustery days, the tents, even connected, can be blown around. So on the days when the wind is heavier, it is important to secure the tent legs to the ground by attaching them to buckets of sand, crates of heavy vegetables, or large rocks. In the past, we have placed the tent legs directly into full crates of beets or parsnips to hold them down, but since the vegetables need to be put on the table to

be sold, this is not a permanent solution. It is good to have a bucket or two filled with sand handy in the van in case a strong wind visits the market.

5. Refer to the Set Up List and Sequence in *French Fries and the Food System* for the rest of the steps involved in set up. (See Attachment: 21: Set Up List and Sequence.)

Attachment 21

Produce Display

The Market

Stands set up
With colorful fruits and veggies,
It seems like the only food that should be eaten.
It's fresh,
Smiling faces loom over it,
And crowds of old black ladies,
Indian housewives with cast marks
And beautiful silken dresses wrapped around them.

-Jesse Lawton, Summer Program Participant, 1997

The produce display is an important element to the aesthetics of the market. It is important to make the most of colors, shapes, and sizes of the vegetables to create an attractive table. For regular customers, it is nice to have a consistent display for the tables so people know where to look for greens, root vegetables, or fruits. As vegetables change through the season, intermix them with the multi-season vegetables. In order to keep the display consistent throughout the season with changing staff, it may be helpful to have a map of the stand which details where vegetables go on the tables. Use the map during the market training so the youth have some idea of the set up before they arrive at the market site.

Produce will arrive from Lincoln and Roxbury in orange harvest crates and white buckets. The trucks usually arrive by 3:15 P.M. Have everyone unload the produce and bring it to the appropriate table. Once the unloading is done, the pairs should begin setting up the table they were assigned during the training. Oversee this carefully to make sure everyone is moving efficiently and placing the vegetables at the correct places.

Once all the produce is unloaded, make sure everyone remembers which table he or she is responsible for setting up. Remind the youth to use the bushel baskets for bigger items of high volume and to use the half-bushel baskets for smaller items. The very small baskets should be used for small things we have a limited quantity of like hot peppers, artichokes and herbs. Encourage the youth to imagine themselves as the customer and to think how the produce will look from the other side of the stand. A typical produce display and the appropriate baskets to use are as follows:

1. Greens Table
   - Collards, lettuce, bok choy, swiss chard in the long metal tubs with about two inches of water for their roots to absorb.
   - Salad mix in a metal tub without water or in a half-bushel basket.
   - Broccoli, cabbage, and peas in half-bushel baskets or in bushel baskets if we have a large quantity.

2. Roots Table – Celery, beets, carrots, daikon, turnips, potatoes, onions, and garlic. If there is a large quantity, use the large bushel basket; otherwise, for a smaller quantity, use the half-bushel basket.

3. Specialty Table – Items like bread, flowers, herbs, beans, honey, okra, and anything else that doesn't have a place on the other tables can go on this center table.

4. Summer or Fruit Crops Table(s) – Tomatillo, peppers, tomatoes, corn, eggplant, cucumbers, summer squash, winter squash, cantaloupes, watermelons. This table may vary depending on availability or to make it easier for customers to see the vegetables on the stand.

Once the produce has been displayed, workers should spray all greens with water. Any overstock of greens should be maintained in the same manner. Squished tomatoes won't sell either, and need to be culled from the stand or overstock. The inventory pair is responsible for restocking and maintaining all vegetables during the market.

Market Signs

Recognizable and eye-catching signs are important for the market. The Food Project has a series of market signs. In addition, the DFA sends out signage that identifies us as a WIC vendor and as selling Massachusetts grown produce. These signs should be posted at the markets. Permanent signage for the market site would be a great asset to the market.

Any advertisement of the market (such as a sandwich board) should indicate the name of the market and site, the city, and dates and times of operation. Multilingual messages ensure that all of our market constituents can become informed. We have sandwich boards that contain these items. There are smaller, blank ones that can be used to advertise specials. Try setting them up a block or two away from the stand and pin up a poster announcing popular vegetables like tomatoes or corn.

A little higher on the left? How's it look now?

Signage that is displayed on the produce stand should have a picture of the vegetable, fruit, or product, some color, multilingual messages, and a place to write the price per pound. This is the recommendation of theDepartment of Food and Agriculture.

Other signage may include harvest dates for certain items, announcements of demonstrations, lists of events, information about The Food Project, and special offers. Be creative and bold!

Take Down

Focus must be held even to the very end of the market when the take down session begins. By the time we are ready to close the market for the day, both youth and staff have been working long hours and are tired. Conditions like this make it very easy for mistakes and frustrations among co-workers to occur. As with the rest of the market experience, the more clearly jobs are given, the more efficiently everyone moves. Below are some steps that will help in finding your own rhythm with closing the market. Note that it helps tremendously to start the closing process before the actual end time of the market. Start half an hour before the market ends.

- Set up a station to weigh the left over produce behind the stand. Place the big scale on an orange crate and begin to line up the overstock on one side of the scale. Assign a few people to consolidate some of the overstock under the tables and transport it to the weighing station.

- Send two people to take down the sandwich board and load it into the van. Have them also take down the banner under the clock, fold it, and put it away.

- Depending on how busy the market is, you may also be able to consolidate the produce on the tables and take down one table or a whole tent while the market is still open. Make sure at least two young people remain behind the stand to help customers while the rest begin to take down the stand.

- At this point the banner behind the stand can be taken down and folded up. If there are not too many customers, have some young people start taking down extra bags and scales. Also, young people can unload boxes from one of the growers' trucks and line them up in rows behind the scale. The overstock will be poured into these boxes to be weighed and then brought to the shelter.

Once closing time arrives, bring the group together to announce the plan of action. When the group is joined together, all focused on a common goal, the work goes faster and is implemented more efficiently.

- The first big job is to get all of the produce off the tables, consolidate the crates of produce, and put same vegetables into boxes to be weighed. The vegetables will then be brought to the shelters. For the most part, only one type of vegetable goes into each box. The Food Project records how well each vegetable sells at the market so it is important to keep the produce separate for weighing and equally critical that weighing is done carefully and written down carefully as well. The grower will either deliver the produce to the shelters in the orange crates, emptying them into containers when they arrive there, or they may have boxes to load the vegetables into so that the vegetables can just be dropped off. Because many of our boxes are donated, they are different sizes and weights. Be sure to explain how to load these boxes to the young people. The sturdier boxes should hold heavier items. Sometimes it is helpful to put heavier items in the smaller boxes to distribute the weight better. The youth can unload the same tables they loaded. They should take the vegetables off the tables and combine them with the overstock into the cardboard boxes.

- Each youth should then bring a box to the scale and the grower will record the weight. They should place the box on the other side of the scale and go back to unloading the table. Eventually, you should have two young people on the other side of the scale, carrying the boxes to the van that will go to the shelter. They should be advised to load the heavier things first, on the bottom.

- Once this is complete, have a few youth clean off the tables, break them down and carry them to the van. Others should consolidate the bins, baskets, and tubs and load them into the van. Four people should break down the tents and bring them to the van.

- You should supervise the loading of the van in order to ensure that there is enough room for everything.

- Towards the end, you should assign someone the job of sweeping the site. We are given the privilege of using the site from the Parks and Recreation Department and it is important that we leave the site looking better than when we found it. The person who sweeps should do a thorough job. Others should check the cement ledge to throw away any garbage we have left there.

- Once this is complete, send the young people home and unload the supplies back into the shed. Often the urban grower is available to help you, but you should make plans for this ahead of time. It is possible that the Volunteer Coordinator could find you a volunteer to help with this task.

Each week, Marcus would get psyched to pack the back of the van as precisely as possible so that each farmers' market table, tent, and bin fit perfectly and could be unloaded with ease at the next market. Over time, he became the Master Conductor of the whole process, coordinating the delivery of infrastructure to the back of the van and then requesting a specific item from the youth who delivered it to him.

"Hand me the table, no, not that table, the longer one. And how about the tents, one, two, where is the third? The pink bin has some heavy stuff in it, be sure that that goes on the bottom and the bin with the hanging scales goes on top. Let's push this one just one more inch to the side and then that sandwich board can fit in right here."

It was at that moment that I realized that for some of these young people, their job wasn't about glamour, or doing the task that would make them a celebrity in the group. It was often about getting the job done, well and with precision, and becoming a master at something. Whether he planned for it or not, Marcus did become a celebrity, for packing the van, like it had never been packed before.

Set up for the market begins around 3:00 pm for a 4:00 pm opening time.

Running the Market Smoothly

"I have learned business skills, people skills, leadership skills, and a lot about community. I have learned to be a motivator without bossing people around. I have learned how to work with people one-on-one."

- Janique Parrott, Farmers' Market Intern, 2001

There is no better way to become a good farmers' market manager than by running a farmers' market. Your skills will only increase with your experience as you find better ways to deal the issues and problems that can come up in this job. The Food Project has run the Farmers' Market at the Dudley Town Common for several years now. Over time, we have learned many things about working in the community, serving customers, and managing the market with young people. Each season, as we become better and better at what we do, we learn more things. Many of the tips we have already gathered have been introduced in early chapters. Below is a sketch of the types of things you can expect at the market and advice on how to make your life and job easier. As stated before, your success at the market will also be based on the personal touch that you bring to the market.

You are off to a good start at the market if the young people staffing the market are prepared, well trained, and on time. The goal is to be open at or before 4:00 P.M. It helps if you are a constant presence with the young people as they staff the market. Set expectations for the young people and hold them to these expectations throughout the market. Structured guidelines of behavior and performance are crucial to success and can provide continuous growth and development for them over the season.

It's An Experience

It is a hot August afternoon. The time is just about 3:45. Most of the vegetables are out of the van and onto the tables and you are running around making sure the seven youth working with you have something to do, that all of the vegetables have been accounted for and that the five customers hovering around the corn are not lifting them into their shopping bags. Welcome to the Farmers' Market at the Dudley Town Common. A wild, fabulous, frustrating and exhilarating experience. It will only get better with each markets you run.

Energy from the young people and the customers can create a frenzied atmosphere at a summer market.

Setting up a beautiful stand efficiently and on time is crucial. But even when it is only three quarters set up, during the height of the season, eight or ten eager shoppers will hover. Though you are running around lifting crates and sweating, customers are anxious to buy despite the fact that you are not ready to sell. After six times of asking folks to please wait and stand back while you try to weave a fifty-pound crate of potatoes around their eager bodies, you may begin to get a little frustrated. One of the most important things you can do for yourself is anticipate this. Know this mantra: *The market can be crazy from 3:45 to 5:00.* It requires a lot of energy from you and the young people. If you prepare yourself for this, you will probably be exhilarated by it.

One way you can ease the hectic nature of the market's opening is by purchasing an "Open/Closed" sign. In the past, we have tried to run a string between orange crates in front of the stand so people would not get close until we were ready. The string we used was a little too thin and unnoticeable and was taken down after we realized it was a liability. You may want to try a combination of a large open/closed sign with thick, brightly colored rope. Just make sure that the site is still accessible to the staff who are setting up.

Customer Service

It was only the third week of the summer program when I began to notice bags of certain vegetables, carefully weighed, stashed neatly beneath the market tables. At first, I assumed they were simply the results of a mistake, that a customer had come and changed their mind and that the youth didn't have time to return the vegetables to the display. So, as the youth workers scurried around me, assisting customers and restocking vegetables, I inconspicuously began to take out the stranded vegetables, putting them back into the bins and baskets that lined the colorful tables. As I was returning the last of the shell beans to their display crate, I heard a young voice call out from the other end of the table, "Kristin! Stop! Those are saved for the shell bean lady!" I immediately knew who "the shell bean lady"

was – and I realized my mistake. The youth had taken an order from one of our regular customers, "the shell bean lady," at the previous market and had put aside a pound for her. This type of attention and care for the customer coupled with an interest in the community and its needs makes The Food Project Market a very special event every week. I think it comes from the fact that the youth, through their work on the land, their service in shelters, their interaction with the community, have developed a deep investment in this market. It is theirs – and they take pride in offering the fruits of their labor to a diverse community that values fresh, delicious produce. The day I was caught returning those shell beans to the table, I learned a great lesson. As a market manager, I am only peripherally in control. I do my best to train the youth in good and efficient market skills. But, in the end, the market's customers were and are in the hands of each of the youth that graces its tables. In this case, the youth had brought customer service to another level, giving special personal attention, and following through with a commitment from market to market. At that moment, the market was busy and bustling, so much so that I couldn't praise the youth for her maturity and care. Instead, I smiled broadly and began carefully the shell beans again.

Getting ready for a fun market day.

We constantly encourage our young people to be as friendly and considerate as possible when helping customers. You have a dual role in this regard. Keep an eye and ear on young people as they help customers, encouraging them to engage customers and go out of their way to help them. You need to be a resource for customers as well. They need to come to you if they have a complaint, problem or question that a young person can't deal with. Anticipate these needs and step into situations as they come up. As the old saying goes, "The customer is always right." Regardless of the situation, it is helpful to keep that in mind in terms of your responses to customers. Granted, we cannot always accommodate our customers, but it should be our starting point. That said, here are some common complaints and situations you may come across during the market season:

- **Complaints about prices.** We price our produce compared to conventional produce and our vegetables are organic, so we feel that we are providing a good product at reasonable prices. We are also willing to look at our prices and if several people complain about the same item, we will revisit the price.

- **Complaints about waiting.** From 4:00 to about 5:00 is usually our busiest time at the market. People know that if they want the more popular items that sell out quickly, they should come as early as they can. For this reason, there can be a lot of impatience and frustration if customers see that we are selling out of corn. This is a time you should step in and help the young people. Also, be aware of customers trying to get a young person to help them while they are helping someone else. You should step in and ask the customer to please wait while the young person helps the initial customer. You may also want to address the crowd, asking for their patience as we try to help everyone as quickly as we can.

- **Theft.** Occasionally a young person will come up to you and say that they think a customer has walked away without paying. You should ask if other youth have seen this and after a couple of validations approach the customer yourself. Ask them if they have paid for their item. Tell them that perhaps they forgot, because we didn't see them pay. This is about all you can do and if the person strongly feels they paid, you should let them go.

- **Disturbance.** The Dudley Town Common is at a highly visible intersection and anyone can walk on the common. Occasionally there will be a drunken person or someone else disturbing the market or the young people. If you see an individual whom you would not feel comfortable approaching, call a community police officer who will come to your assistance. You should try to anticipate these moments and stop them before they occur. Often, ignoring these people rather than confronting them will make them go away. If they are really disturbing the youth, ask if you could speak to them for a second away from the group. Tell them that we are trying to train young people here and you would appreciate it if they let the young people focus on their job.

Maintaining Relationships with Local Businesses

It is important for any existing business to create positive relationships with its neighbors. Shops, stores, and outlets can foster business for each other. The businesses that line the streets across from the Dudley Town Common are important stakeholders in our market. The Food Project youth will often use their services and buy their products. Many of their customers are drawn to the colorful tents and beautiful vegetables displayed on the square. On the other hand,

The Food Project has learned through surveys that the presence of the market has increased business throughout the area. The neighboring businesses often come over to buy our produce and talk to the youth. At market's end, we frequently give some of the produce to the store owners and workers.

For these reasons and others, these are important relationships for the market manager to foster. When you put up posters around the neighborhood, use that time to introduce yourself to the business owners and let them know of any changes to the market for the season. In the last couple of years, there has been some turn around in the local businesses, so some of the current ones may not be aware of the market.

So much to choose from!

Below is a list of our recent business relationships. Explore others to establish new connections in the community.

- There is a pizza place across the street from the market. During the summer, the youth often take their dinner break there. In the past, the owners have bought tomatoes and other vegetables from the market for their restaurant.
- Also across the street are two places that allow us to use their bathrooms- Montesino's Tire Shop and La Alianza Hispana. Both also allow us to advertise for the market with flyers and signs.
- Montesino's has been receptive to us in times when we needed tools or a hand in fixing something.
- La Plazita, a small store across the street, serves good lemonade that we've enjoyed over the years.

It is important to The Food Project to make sure that our effect in the neighborhood and at the market site is a positive one. Try checking in with the businesses during the season and at the end. See how things went for them. Was there anything about the market that disrupted their business? Would they like to see any changes to the market? Would they like to advertise with us or become more involved with The Food Project or the market itself?

Events at the Market

During the planning section of this manual, we advise you to develop an outreach plan for the season. While our central mission for the market is to provide produce to the community at reasonable prices, we also aim to be a resource to the community and try to offer more than just vegetables. This distinguishes us from a supermarket and can be an incentive for customers to shop with us; they see the positive impact we can have on the community as well as on individuals. Additional vendors or events occurring at the market can encourage customers to linger, buy more vegetables, get connected with their neighbors and have an enjoyable shopping experience.

A booth from the University of Massachusetts Extension program participating at The Food Project's Farmers' Market.

As stated earlier, the amount of time you spend on outreach will depend on the amount of time available to be dedicated to the market outside the actual market operations. At the very least, it is important to celebrate the first market and the final market. Also, it is nice to have other market festivals throughout the season. This helps to bring more people to the market and to keep momentum up during the season. As stated earlier, you could do a big celebration mid-season or a few smaller celebrations around the height of the harvest season of certain popular vegetables. Talk with the public outreach coordinator at the organization to see if they might like to collaborate on an outreach event at the market or if they would be willing to offer their help as you develop and coordinate events.

Here are some ideas and themes to think about as you plan events. Who is your audience? What are the types of things customers in general as well as this specific community enjoy? Free food is always a sure fire way to generate happiness at an event. In the past we have celebrated the first market by giving away strawberries with free strawberry shortcake. Set up a separate table, either with a capable young person or a staff member running that table throughout the market. We have a good supply of kitchen items for this. Be sure to check in with the growers about which vegetables will be in surplus this season. Music is another item that will attract people to your celebrations. In the past, we have had a local Cape Verdean string band play at the market. A Roxbury group called the Silver Leaf Gospel Singers set up on the ledge behind the stand and played for an hour. We had a table with one of our growers' assistants making corncob dolls with children at the market. At another table a staff member handed out cookies for people to decorate with sprinkles and icing as well as candied apples for a cheap price and popcorn. At this last market, we were also able to buy apples from a local orchard and resell them.

Allow yourself at least a month to plan an event such as the final market celebration described above. This will allow you some time to play phone tag with possible presenters or vendors. Inventory staff talents at a staff meeting and ask if anyone would be interested in contributing to a market celebration. Using the products that are in season at the time of the celebration is a great way to use extra produce, promote a seasonal product and to teach customers how to use that item. You may want to feature a vegetable during the season like tomatoes and have a visiting chef share five different ways to prepare a tomato. For the children, maybe you could have someone dress up in a tomato suit and hand out seeds. Be creative and also be reasonable. Don't plan an enormous event if you don't have the time to really do it or the staff to support you. Many great events can come off low-cost and without a ton of planning.

In addition, always remember who your audience is. What types of music do they listen to, what do they like to eat? In this neighborhood, Cape Verdean music, gospel, hip-hop, R&B are all popular types of music. As you go out to lunch in the neighborhood, you will notice the popular types of food; rice and beans, pork, barbecued chicken, collards cooked with meat, mashed potatoes and biscuits. It will always be hard to convince someone to try something completely different than what they are used to. Take what people already know and add one of our vegetables to it and it will be more easily received.

Musicians playing during a summer market.

Make sure you check your budget for events and food and work within that total when buying food or hiring entertainment for these celebrations.

Keeping Business Records

"My overall expectation for my position was to work with the market manager on topics concerning the market. I was interested in doing all the things that involved running a market. For example, trainings, analyzing, finding prices and so on. I expected it to be demanding in terms of there being a lot of work and there was."

- Janique Parrott, Farmers' Market Intern, 2001

Keeping track of details, exchanges, and customers for any business is important. At The Food Project, the Farmers' Market is an enterprise as well as a community outreach and education program. In order to make the most of our enterprise effort and to improve the market year to year, it is greatly beneficial to have numbers and observations to assist and guide decision-making. In addition to the task of keeping accurate records, The Food Project seeks to carry out data collection and record keeping in partnership with youth. It is especially important that the market manager is on top of the record keeping process and system because it can get confusing, especially when youth are only working two or three of the markets. To mix managing young people with managing a business requires a high level of organization and patience for the market manager. Make sure you know the systems well before introducing them to the youth. When training the youth on record keeping at the market, make sure they understand the larger context for what they are doing: Why are we keeping these records? How is data like this helpful? What skills will they be developing when weighing, counting, and recording at the market? In the course of your tenure as the market manager, you may come up with more effective ways to record and organize the data that you are collecting. Before changing any of the templates for the market data, share these ideas with your supervisor and the business manager so they make suggestions and assist you in changing the present system.

Record Keeping Tools

Something to carry your record keeping materials is essential. You may want to keep a market manager's notebook with you. At The Food Project, market managers have often carried a plastic file box. They are waterproof and sturdy and enable you to carry multiple files at a time. In this box, have a small three ring binder to carry your record, two file folders, a file with the masters of the recipes, and a section to keep coupons and cash to be deposited. Additional folders or organizational tools may be required for your own style, but use these as a guide.

Record Keeping with Youth

Record keeping at The Food Project demands attention to detail. Over the course of the years, however, systems have been set up that include spreadsheets and templates that assist in organizing the information. As long as time is invested, keeping accurate records need not be a difficult aspect of the market. The Food Project keeps careful records of where its produce goes – whether to markets, shelters, enterprise, or our Community Supported Agriculture Program. For our own organizational use and for information for our funders, we need to accurately report the poundage of vegetables that goes through each distribution channel.

After a crew has run two markets, the market manager will lead them in an Accounting lesson, a mini-Market analysis. See *French Fries and the Food System* for an explanation of this lesson.

Keeping Invoices. When the growers from Lincoln and Roxbury arrive at the market site with the harvested produce, each will have an invoice for the vegetables they are delivering. The market manager is given the yellow copy of this invoice. The information will be transferred from the invoice to another spreadsheet. However, it is necessary to save the invoices in a folder in case of the possibility that one of the numbers is copied wrong and there is a discrepancy that occurs in the records.

Recording the vegetable weights on spreadsheets. The vegetable weights from the invoices of both farms are transferred onto the vegetable sales logs (see Attachments 22-24: Market Sales Logs). Each week, the manager should make two copies of the sales logs for each of the markets. The first copy can be used as a guide when the manager or intern goes to the supermarket to check prices at the beginning of every week. In addition, the sheet can be used to refer to when speaking with the growers about the quantity and type of vegetables that they will bring to the market. The second copy is used at the market. Using the invoice measurements, a responsible youth should be chosen to transfer the information onto the sales log sheet. The first two columns, L/Quantity and R/Quantity are reserved for jotting down the quantity estimated for delivery to market by the Lincoln grower and the Roxbury grower. The next two columns are for actual weights from Lincoln and Roxbury. When the grower arrives with the vegetables, they will hand the manager the invoice that details each vegetable, the quantity of each and the total in price and pounds. You can easily explain to a young person how to copy this information into the book during the market. With this information, we know how much of each vegetable came into the market, from Roxbury and Lincoln, as well as the poundage of all the vegetables together.

Attachment 22
Attachment 23
Attachment 24

Young people testing their market skills.

Recording end weights. At the end of the market, all of the produce is weighed before being delivered to the shelters. The urban grower and a selected youth will weigh all the left over produce. The grower fills out an invoice for the shelter delivery that itemizes each vegetable and its poundage. A copy of this invoice is given to the shelter and the other is kept by the grower for the files. As the grower fills out the invoice with weights, a responsible youth fills in the Sales Log sheet with the same weights. When weighing the vegetables to go to shelters, the youth should use the categories farther to the right on the sheet, marked Lincoln and Roxbury. Right across the row, the youth are able to compare how much of one vegetable came in at the beginning of the market, how much was sold, and how much is being donated to the shelters.

Revenue record keeping. Cash and coupon documentation at the end of each market is another record keeping necessity. It is important to keep track of the revenue for each specific market so we can track our progress from year to year. After the market is finished, make sure that the cash register or cash box makes it safely to the van and is taken to the office once the vehicle is packed. During the Accounting sessions at the end of the week, each crew will count up the total cash and coupons received for the markets they ran. In order to do this, the Market Manager should print out copies of the Master Accounting Sheet (See Attachment 25). This will provide a worksheet for the youth to calculate the totals for the market. The sheet allows you to record cash, coupons, and other costs while figuring out your profit for the market. It also has a place to record the numbers of customers. This spreadsheet is in the accounting folder in the Farmers' Market folder on Shared Files, so you can transfer the numbers to the computer if it helps you keep track of your records. Save the Market Accounting sheets to transfer the information into the master market spreadsheet. Depending on the amount of help you have, you may want to fill these sheets out for each market during the academic year as well. If you have a market intern, they could also be trained to do this .

Conducting surveys in the community helps The Food Project run a better farmers' market.

Customer Tally. Over the years, The Food Project has been building a customer base for the market. Keeping records of customers allows us to gauge how much we are growing each year. The numbers of customers at markets during August and September are often much greater than the other months of the market and it is good to use predictions to get a sense of how much produce we need for each market. Customer tallies have been taken in a variety of ways. In the past, each individual youth was responsible for making receipt slips for the customers they worked with. At the end of the market, these slips would be counted to determine the customers that had come that day. Theyear we had a cash register at the end of the market, we used the receipt slip to count the number of transactions we had over the course of the three hours. Another method that has been used is to have a customer tally sheet next to the cash box or register and instruct youth workers to put a check mark on the sheet each time they sell to a customer. Each of these methods has its strengths and weaknesses and none of them are a completely accurate method of recording. In the early stages of planning the market, decide which method seems most efficient to you and work to make it as accurate as possible throughout the season.

Compiling Data on the Market Spreadsheet

Spreadsheets are an easy way to keep track of the variety of information collected about the market. During the 2000 season, a new series of linked spreadsheets were developed for the market manager to keep track of all the information necessary for the quarterly reports (see Attachments 26-32). For managers outside of The Food Project, these forms can be used as a guide and altered according to the specific needs of the market. **Attachments 26-32**

For The Food Project market managers, these spreadsheets are on shared files and should be used to accurately record the happenings at the market throughout the season. Ultimately, all of the information described above – vegetable weights, revenue, and customers - will need to be inputted to this spreadsheet. The market manager will save a lot of time at the end of the season by reserving a weekly or biweekly time to do the data entry necessary for market records. By following the earlier recommendation of cashing in the coupons on a biweekly basis, the manager can use this time to input the quantities into the market spreadsheet. In order to do this, the manager needs to empty the cash register after each market and organize the cash and the coupons; label the coupons with the Certification Number, and then record them into the spreadsheet.

The market spreadsheet is a series of eight linked sheets. When filling out the spreadsheet, begin with the last sheet called 'Farm % Breakdown by #'s'(Attachment 32) and 'Farm % Breakdown by Dollars' (Attachment 31). When you input numbers into the market spreadsheet only input the data entry into the columns with dashed marks outlining them. The rest of the columns contain pre-programmed commands. You will see that as you input numbers into the spreadsheet, it will automatically carry those numbers over to certain spreadsheets and calculate them. Input the last two spreadsheets *first* with the information you get from the growers on the total quantity and price of the produce they brought to the market. The next sheet is titled 'Market Donation Sheet' (Attachment 30). Input the total end weight at the market and that is the amount that will be donated. The next two sheets titled 'Boston Stats' (Attachment 29) and 'Lincoln Stats' (Attachment 28) and do not have any data entry. The final sheet for data entry is called 'Market Receipts' (Attachment 27) and that is where you input the total cash, coupons and customers for that market. When all the figures from these six sheets are inputted, the final spreadsheet, the Market Statistical Summary Spreadsheet (Attachment 26) is completed. Be careful to check all information to determine whether the formulas have translated a correct answer. One small error in data input can alter the records significantly.

Attachment 32
Attachment 31
Attachment 30
Attachment 29
Attachment 28
Attachment 27
Attachment 26

Evaluation

- *Completing Your Records*
- *Analyzing the Market*
- *The Marketplace*

End of the year evaluations are crucial for each program within The Food Project. They allow for definitive completion of tasks and formal evaluation. Most importantly, they enable us to develop tools and strategies to improve the next season. At the end of your market season, you will have several tools available with which to evaluate the market.

Completing Your Records

The final records for the Farmers' Market are needed for the rural agriculture and urban agriculture year-end reports. All of the information needed should be found on the market spreadsheet. At the end of the market season, go through the spreadsheet and make sure all of the information is filled in accurately. Notice if the functions on the spreadsheet have worked. If things are highlighted in red or there are asterisks in the cells instead of numbers, this could signify an error. Sometimes the solution is as simple as enlarging the width of the cell to make room for all of the numbers. The red could be telling you that there is a high margin of error in the data. It is helpful to look at these numbers because it could mean there was an error in your data entry or at the market. If you are having trouble deciphering the spread sheet, speak with our business manager, your supervisor, or one of the growers.

At this time, you should also make sure that you have sent all of the coupons into the Federation of Massachusetts Farmers' Markets for redemption. They will send you a mailing in November letting you know when to expect the check. This check should go directly to the business manager and a copy should go into your records.

Analyzing the Market

So, they must have really listened. To the growers, to the neighbor gardeners, to their customers, to the Urban Education Interns, to each other, and to the land. They listened in workshops, in the field, in the shelters, at the market site. When the DIRT Crew got up to present the results of the market analysis this fall, they did not merely de-

scribe numbers, graphs, and charts. They took the information they had synthesized and compiled, added it to the body of knowledge they had built over the five months of involvement with The Food Project, and shared and reflected with strong voices and an ability to add depth and nuance to their answers.

Did our customer base decrease in fall of 2001? The line graphs demonstrated that our customer base was lower compared to 2000 during the final two months of the market. How could this be explained? Carmen bent down over the graph, placing her finger on one point along the line, and named the spot – "Here is September 11th." From the 11th on, we all observed the gradual, but noticeable, decline of customers and revenue at the market. There were certainly other factors to be considered, in particular that coupons were already distributed and available for people regardless of the economic downturn, but it was obvious that the tragedy of September 11th had stretched to have its own affect on Dudley Town Common.

How much can greens weigh?

What are our vegetables really worth? The monetary value of the vegetables we sold at the market this season was greater than the actual money that we collected in cash and coupons. Why was this so? Shatara explained, "The growers value the vegetables at a higher price, the real cost of organic vegetables. When we sell the produce at the market, we sell it for less than its really worth, so that people can afford it."

Warren, Ibe, and Alex described the vegetable stats for the season. After practicing his part studiously, Ibe confidently described the percentage of each vegetable that was sold over the season and the remaining percentage that was donated. "Is this a good measure of vegetable popularity?" The youth were asked, "Aren't there other factors to be considered?" Warren replied, "The summer vegetables sell better partially because there are more people at the market during those weeks." "Yes," Alex explained. "But the corn attracts people to the market. We have more people because there are popular vegetables." Suggestions arose as to how we could more accurately measure vegetable salability. We had only taken one of many steps toward a more thorough analysis of our market vegetables, but that step had been taken nonetheless.

Michelle, Emma, and Jen described what our customers had to say about the market. A question came from the audience, "What do you say to a person who argues that organic produce should be less ex-

pensive than conventional because you don't have to add anything to it? To grow conventional vegetables, you need to buy inputs – chemicals, pesticides, fertilizers." Michelle didn't skip a beat, "I would tell them that it actually takes more work and effort and money to grow food organically. We can't just kill bugs and diseases with chemicals so we need to feed our soil and pick off bugs to make sure that our plants survive. It takes more time and more careful planning and more financial investment to be organic farmers."

Questions were actively shot out and most of the DIRT Crew Members, sitting around the circle, were excited to interject their opinions based in observation, experience, and an intimate study of the "numbers." By the time Derry, Jose, and Katie stood up to present the Youth Analysis, the discussion of what the DIRT Crew thought and felt had been already been thoroughly touched upon. Although they had spent two weeks in the summer running the market, five sessions preparing the presentation, many afternoons of compiling information and creating graphs, the market was still alive for them. It was obvious in the Analysis Presentation that the market had not been archived in the DIRT Crew's minds through graphing and charting quantitative data, but that it had remained vibrant and alive with many colorful human faces, lively interactions, and a critical awareness of both the frustrating and effective systems of its operation.

A couple of different analyses will happen towards the end of the market season. At the end of the summer program, there will be an evaluation of all aspects of the program by the staff and leaders. There will be a section on the Farmers' Market and you should make a note to yourself before the analysis if there is anything specific about the market you would like to have feedback on. This could include the youth training if you added anything new, the organization of workers at the market, and so on. Be sure to get a copy of this evaluation for your records.

Youth Analysis

Another evaluation occurs with the youth that have committed to the Academic Year Program. The Farmers' Market analysis in the fall is a major part of their curriculum. See *French Fries and the Food System* for the lesson plans for the analysis and see below for a brief description of the elements involved in the market analysis.

Before planning the fall analysis, however, consider the results from the past years and how these have informed future markets. Considering the many elements of the market is a lot of work and time. Past years' analyses have focused broadly at all the elements of the market and this broad analysis has been conducted at the expense of a more detailed study of particular elements. As we become better at running the market, and improving it comes from more detailed tweaking, it may become important to pick a specific aspect of the market and thoroughly analyze that for the year instead of trying to look at everything. If we could do a deeper study on advertising and marketing, then we could institute a more effective marketing strategy in the future.

Developing a Market Survey. A central part of the fall analysis is surveying the customers at the market. It is important to begin the surveys as early as possible in the fall to get as many returns as possible and to improve your results. Arrange with the Academic Year Program Coordinator when you can do the first lesson of the Market Analysis with the DIRT crew. This first lesson is focused on creating a survey for the market. It is helpful for the manager to have reviewed past years' surveys and results to have an overall sense of the successes and challenges to be learned from. With the youth, however, it is best to begin by brainstorming about what would be interesting and useful to know from our customers. After compiling questions from the brainstorm, have the youth look through the past surveys and see if they missed anything that might be important to their analysis. Do not stick to the format, however, if you see a new need or a different direction in which the survey should go.

Sunflowers grown in The Food Project's greenhouse for sale on our city lot.

Conducting the Survey. The manager will also discuss the methods with which to deliver the market survey (lesson described in *French Fries and the Food System*). When planning for the number of youth needed to run a fall market, keep in mind that one or two of the youth should be reserved for giving surveys to customers. A past method has been to have at least one person doing a survey the whole market with two people during the 4:00 p.m. rush. Halfway through the market, have the survey person switch with someone who is operating the market. To provide incentives for the youth, try setting up a competition for who can do the most surveys in a season. Aim for at least 100 surveys by the end of the market season.

Analyzing the Survey. When the survey is completed and the market has finished for the season, the main part of the market analysis is conducted. The youth gather the information collected on the surveys and create charts and graphs to illustrate the results. The objective is to use customer suggestions, needs, and desires to influence how we run the market. We need to know convenient market times for people, what products customers are looking for, what we could do to improve our service, and many other aspects. We have learned from previous years that it is important to have the survey available in several different languages for all of our customers. Additionally, it may also be advantageous to offer the survey in the summer months as well as the autumn to get a better idea of what the majority of our customers want.

Vegetable Analysis. During the market analysis, the youth take time looking at the vegetable log sheets that they filled out during the season and the spreadsheets that were created to compile all the information. From these, they graph the amount of poundage coming in from Lincoln versus that from Boston. They also determine how much of the produce was sold at the market versus what was donated to the shelters. Last year, the youth conducted a more specific analysis of which vegetables sold best by tracing each vegetable over the course of the season to see what sold best. There are many possibilities for deepening and expanding the current vegetable analysis. The growers might have some input as to information that would be useful for them to have.

Youth Experience. At the end of each week during the summer program, the youth crews analyze the two markets that they have run. As has been already discussed, part of their analysis is quantitative - counting and recording cash, coupons, and customers. The other part, however, is qualitative, when they reflect on the market systems from their perspective. Did the market run smoothly? What parts needed improvement? How was working in pairs? Was there a backlog at the cash register? How can we improve our customer service? (See Attachment 33: Market Analysis Workshop and Attachment 25: Accounting Sheet.) Each crew's reflections are recorded on a flip chart page to be referred to later. In the fall, the youth in the Academic Year Program (the DIRT crew) conducts its own qualitative analysis of the autumn markets. The DIRT crew reviews the reflections from the seven crews that worked the markets during the summer. Through compiling their own responses and the reflections of the summer, the DIRT crew comes up with suggestions for next year's market.

Attachment 33
Attachment 25

Preparing the Market Analysis Presentation. All the steps above are focused on creating a final presentation that the young people give to the growers, interested staff, and board members and friends. During the weeks of the lessons, schedule the DIRT Crew on afternoons during the week to finish the work not completed during the sessions themselves. In preparing for the overall presentation, the youth should have access to all the market spreadsheets. These spreadsheets are the fullest quantitative analysis for the market season. Spend a session preparing the youth for the presentation. (Refer to *Growing Together* for guidance on training youth in public speaking and presentations.)

The Market Analysis Presentation. The final presentation of the market analysis is an event and should be treated accordingly. Every year the market runs from June through October and has lasting impacts on the youth, the community, and the neighborhood. The youth have harvested vegetables, sold them, and then spent many hours reflecting on the successes and challenges of the market. With all the effort, thought, and time, it is critical to plan the presentation to reflect the importance of what the youth have created. In past years, we have begun or followed the presentation with a shared meal or other refreshments. The Food Project seeks to involve as many of our staff, board members, and funders as possible in the final event. By bringing people together for the analysis, we are able to gather input both on the results of the analysis and how we came to them. In addition, with an engaged audience, the youth see the importance of their work in the future of the market. Taking pictures or filming the presentation can create a valuable resource for future market managers or for use at conferences. The presentation and all of the tools used throughout the process of the analysis should be used when developing the annual plan for the next year and adding new initiatives to the Farmers' Market.

The Marketplace

It was the day after the Dudley Street Neighborhood Initiative (DSNI) Board Election. The DSNI Staff, always great supporters of The Food Project's Farmers' Market, arrived at the market for their regular visit. Strawberries, snap peas ("just put them in a little bit of water and then right in the microwave...they are excellent!"), a head

of lettuce, a bunch of beets. They stood, chatting with the youth, asking questions about what vegetable would be ripening next. "When will tomatoes and corn be brightening your market tables?" During the lively conversation, several neighborhood women arrived from St. Patrick's parish. While deciding between zucchini and patty pan squash ("is one sweeter than the other?"), both DSNI staff and the St. Patrick's entourage reflected upon the event of the night before – the impressive turnout, the delicious food, and the quality of the Board candidates.

From behind The Food Project market table, selling her calaloo, mint, and irises, Sylvia, our neighbor gardener, chimed in to the discussion. She began with stories from a different time, stretching back to Dudley neighborhood events of 10, 20, 30 years ago. We all listened attentively to her words, to her particular rendition of the history of the neighborhood, to her account of the Dudley story. DSNI staff, neighbors, and The Food Project youth stood quietly, somehow transfixed by the humility and insight of the true tale that she was weaving for us, a story that spoke to the uniqueness of the place where we all stood, the land that we gardened, the people who worked and lived around us, the narrow one-way streets that created the map of the Dudley neighborhood. On that day, in that conversation, in the midst of fresh, organic vegetables and open minds, time stopped for a moment.

Only half way through Sylvia's story did I realize that the exchange before me was both planned and happenstance. That year, The Food Project Market on Dudley Town Common was in its fifth year. From June – October on Tuesdays and Thursdays for all those years, youth and staff harvested vegetables and fruits, set up tents and tables, offered fresh produce and resources, and greeted customers. Each season, we plan, organize, and prepare to carry out this event over 40 times in five months. We do not just happen to be there – we commit to being there, rain or shine, in sweltering heat or on chilly autumn days.

The DSNI staff did not just happen to be there that day – they come by the market nearly every week. They arrive early to get the best variety. They introduce themselves to the youth, and linger to chat with residents and visitors. The neighborhood women from St. Patrick's planned to see the DSNI crew at the market to review the event of the night before. They had agreed on the market as a "meeting place" to discuss neighborhood happenings and politics.

Sylvia, our historian, storyteller, and gardener, didn't just happen to be at the market either. Multiple years of cultivating relationships has facilitated connections between The Food Project, neighborhood gardeners, and the land that has enabled residents to sell their harvest to the community, enriching everyone's experience of community and knowledge of farming in urban areas.

Sampling tasty treats at a market celebration.

The pieces were in place – a park permit, clean and vibrant soil, beautiful vegetables, an involved and dedicated neighborhood organization, active and concerned neighbors, and interested and motivated youth workers. Yet, conversations such as the one between DSNI staff, St. Patrick's parishioners, neighborhood gardeners, and The Food Project staff and youth, could never be fully orchestrated. Exchanges such as these that broaden minds to the history of place and struggle, to change and progress, to evolution and community are unpredictable and brilliant gifts. And for us at The Food Project, these conversations that take place near the vegetables that we so carefully cultivate, with the youth that we so firmly believe in, in a community that we so excitedly participate in, these are the blessings of the market place.

How is our Farmers' Market an expression of our vision – personal and social change through sustainable agriculture? Certainly, the market challenges each one of the youth personally - to develop skills of presentation, customer service, communication, and teamwork. But is our market also about social change? Historically, societies that have engaged in vibrant trading of crops at market places have been more likely to delve into the broader fruits of cultural exchange – the sharing of languages, culture, religion, literature, and worldviews. While fruits and vegetables were being thrown from ships to wheelbarrows to be carted to homes around the world, agreements were formed, wars waged, ideas exchanged, friendships made, minds stretched and perspectives changed forever. Societies were and are changed by the interactions that happen in the market place – especially when conversations are sparked by thoughtful, committed, productive individuals with differing perspectives but similar goals.

The Food Project market on Dudley Town Common run by youth and adults committed to sustainable growing, environmental stewardship, community involvement and service follows in the ancient market tradition. Here, in this sunny spot in Roxbury, among people from around the world, an elderly woman tells her story of this particular place. That conversation - Sylvia's words, the youth's questions, the DSNI staff's insights, the neighbors' listening ears – in its simplicity and directness, amid a table overflowing with color and health, has the effect of both widening perspectives and shifting thinking, setting down the seeds for the powerful connections that do nothing less than lead to true transformation of people, neighborhoods, and societies.

Directions to the Dudley Town Common Farmers' Market

From Ruggles MBTA Station or Dudley Station:

Take Bus 15 to Upham's Corner or Bus 45 to Blue Hill Ave. Both buses go directly by the market, which is on the corner of Blue Hill Avenue and Dudley St. Landmarks include a large circular clock in front of the market site, Nuestra Communidad, and St. Patrick's Church. If you are on Bus 15, you get off on Dudley St. and if you are on Bus 45, you will get off on Blue Hill Ave.

From Upham's Corner:

Take 15 toward Dudley Station and get off the bus at the intersection of Blue Hill Ave and Dudley St. The market site is on your left.

From Mass Pike (Rt. 90):

Take Mass Pike East all the way under Boston until it connects to Interstate 93. Take 93 South, then:

From 93S/Fitzgerald Expressway:

Take Exit 18 Mass Ave/Roxbury. At the bottom of the Exit, turn right and continue straight to light at Mass Ave. Turn left onto Mass Ave (gas station on your right.) Then:

From Mass Ave.

Take Mass Ave south from Boston (away from Charles river). Continue past the turn off to 93 Expressway, stay to right at blinking yellow light. Turn right at next light (Magazine St.). Go straight on Magazine St. until it dead ends at Dudley Street. The Farmers' Market will be right in front of you at the corner of Dudley St. and Blue Hill Ave.

Attachment 2

Dexter Miller
Boston Parks and Recreation
1010 Massachusetts Avenue, Third Floor
Roxbury, MA 02118

Dear Dexter Miller:

I am writing to request a permit from the Boston Park and Recreation Commission for the use of park space at Dudley Town Commons, Roxbury (Blue Hill Ave. and Dudley Street). For the past several years you have granted The Food Project use of this space for our Farmers' Market held from June until late October. We have obtained approval from the Dudley Town Commons Advisory Committee (contact individual: Liza Veras) to operate the Market on this site on Tuesdays and Thursdays from 3:30 to 7:30 (this includes setup and take down- actual mkt. operation is from 4-7pm) and have been included in their Park use plan.

Last year was another successful year for our Farmers' Market with the inclusion of another farmer, Drumlin Farm, which added variety and vitality to the market operation. The successful use of this park space has been evident to all in the community and is reflected in the number of residents we serve. In the summer of 1999, we provided fresh vegetables and fruits to a total of 1700 customers at our Thursday market. We look forward to offering high quality locally grown vegetables and fruits to Roxbury residents at the same site again this year.

Produce for sale at the Market comes from The Food Project's Summer Youth Program where fifty teens from Roxbury and greater Boston grow vegetables on farms inside and outside the city. This year in addition to produce, customers will have access to nutrition information as well as organic gardening information.

Nearby businesses have expressed their support for the Market and say they see increased business activity during its operation. We hope to continue this mutually beneficial relationship in the community.

We look forward to working with the Parks Department to make the Market a success in 2000. We welcome your questions and guidance about the use of this space and as we get closer, I will send over some flyers and information for your office.

Sincerely,

Sara Coblyn
Market Manager

Attachment 3

June 10, 1999

Joseph Kelly
City of Boston, Dept. of Transportation
City Hall, Rm. 721
Boston, MA 02201

Dear Mr. Kelly,

Thank you for speaking with me yesterday regarding the use of Parking Lot #27, Blue Hill Ave., Roxbury. I am following up with a written request to utilize this parking lot for our Farmers' Market, each Thursday afternoon (June-October), from 4:00 PM to 7:00 PM.

In the past we have had difficulty accommodating customers who drive by the market, due to the high need for parking in this business area. I am unsure of the current use of Lot #27 during the high traffic hours, but assume that it provides customers with the parking they need to patron area businesses. These relationships with area businesses are extremely important to us and vital to the future use of park space in this area for our farmers' market. Throughout this process, we would hope to maintain these relationships.

We hope that sharing or full use of this lot would allow us to serve more customers and add to the overall patronage of all stores. We currently have a "Parking" sandwich board and cones that could be used to designate the parking area for market parking. If your department requires any other signage or structures, we would be more than glad to follow your guidelines. Thank you again for your time. I look forward to speaking with you soon regarding this request.

Regards,

Colleen O'Brien, Market Manager
The Food Project, P.O. Box 705, Lincoln, MA 01773
Phone: 617.442.1322 or Voicemail: 781.259.8621

Attachment 4

Farmers' Market Inventory

Name of Item	Current #	Optimal #	Current Condition	Maintenance Recommendation	Intials
Bags, clear produce	low	1 lug full		Check budget to purchase in bulk/donations	Baheem
Bags, handle	low	2 lugs full		Check budget/request donations	
Banner	2	2	ok	check for decal that is not sticking	
Baskets-1/2 bushel	8	10	ok	Reorder through New England Basket Co.	
Baskets-bushel	10	10	ok	Reorder through New England Basket Co.	
Baskets-random	12	10	ok		
Metal Bins	4	5	one is broken	Buy a couple more to display veggies-wash and check	
Broom/Dustpan	2	2	ok	You may want to locate push broom	
Calculator	2	2		For use during accounting lesson	
Calculator w/tape	0	8		For use with cash register- one per youth	
Cash box	2	2	ok		
Cash register	0	2		Need to go over cash register plan with business manager	
Chef Demonstration	Supplied by Chef				
Cups, water	0	Couple Packs		Encourage youth to use water bottles	
Dry Erase Cleaner	0	Couple bottles			
Duct tape	half roll	4			
Extension Cord-orange	2	2		Check with urban ag.	
Hanging Scale dials	5	5	plastic shield/glass broken	Check rim- okay for glass to be missing for	
Hot plate	1	1	good	consistently check to make sure works	
Markers, Dry Erase	0	2			
Markers, Permanent	0	20			
Nightly Deposit items	0			Set up if necessary	
Notebook	Kept on site			Contains sales logs, accounting sheets, invoices	
Paint Stirrer	40+	60+	Used for Produce Tags, ok		
Paper towels	half roll	several rolls			
pens/pencils	10	100			
Produce Tags	40+	60+	dirty, missing vegetables	Make new ones for all produce, bilingual	
Receipt pin	1	2		For holding receipts, also can put them under cash drawer	
Rubber bands	half bag	2 bags		For bunching veggies, flowers	
Sales bibs	4	15	ok	Need to be washed/bought at Home Depot	
Sales books	whole box	a box		Keep an eye on this and reorder with office manager	
Sandwich Board, Little	1	1 or 2	okay	Little one- can be used to pin a sign onto it	
Sandwich Board, Big	2	2	ok	needs repainting	
Scale-100 lb. 4 oz	1	1		Hold by bottom or dial will pop out	
Scissors	2	2			
Scoops	5	5	ok	Assign each scoop to dial for inspection	
Signs	2	2 or 3	FP Mission & Availability	Board is broken/needs new laminate	
Small stool	1	1	ok	Needs to be scrubbed, used for elders	
Spray bottles	4	5	ok	Clean/Buy at Home Depot	
Staple Gun/staples	1	1	ok		
Stapler/staples	1	1			
Tablecloths	2	4	ok	Depends on table size	
Tables, Small	8	8	3 in good repair	Try to fix or trash and just use longer tables	
Tent, Repair Kit	1 bag	1 bag			
Tents	4	4	ok	Two white tents sturdy, use first, other two work- gentle	
Tools, general	2	2		Hammer, screwdriver, wrench, plyers	
Trash liners	1	2			
Twine	1	2			
Water jug	1	1 or 2	ok	One at market good mostly, two in hot weather	
Wooden Veggie Bin	1	1	ok	Replace plastic liner/screw in divider	
	c				

Chris Mesarch
Fresh Pond Bread & Circus
200 Alewife Brook Pkwy.
Cambridge, MA 02138

Dear Chris,

Thanks for speaking with me regarding the potential donation of some produce and grocery bags for the Dudley Town Common and Codman Square Food Project Farmers' Markets. I am following up on our conversation with this written request to clarify some of the supply needs around the markets.

In the past we have received bag donations from a variety of local markets in the area. We have not given special attention to relationship building around these requests because we felt lucky to get whatever they had to offer us. My newest commitment, as Market Manager, is to share our mission with donors a little more, so that they can understand just how valuable and special their donation is to our programming and enterprise operations. It has been great working with folks at Bread & Circus because we have similar commitments to our customers, farmers, and communities. It would be a treat to send customers home with fresh vegetables in Bread & Circus bags!

We have had two types of bags at our markets in the past: the clear produce bags that come in rolls and the plastic bags with handles. Any bags would help the cause, but these are the easiest to handle and the most practical for our market. The farmers' market begins June 17th and will continue until October, but I expect to ask for donations from several other stores, if you are unable to supply us with a large supply of bags. Any donation that seems feasible for Bread & Circus is great for us!

Thanks for your consideration, Chris, and your continued commitment to us. Everyone at The Food Project appreciates your efforts. Happy late spring! If you have any interest in visiting a market, please let me know...we could always use professional feedback on our business! I will follow up soon.

Regards,

Colleen O'Brien
Market Manager, 617.442-1322 or voicemail 781.259.8621.

Attachment 6-1

1999 Dudley Town Commons Farmers' Market

Site use for Dudley Town Common, South Roxbury, Massachusetts

<u>Market Organizer</u>

The Food Project, Inc.
Contact: Colleen O'Brien 781.259.8621 (x20)

<u>Neighborhood Advisory Committee</u>

DSNI, Dudley Town Commons Advisory Committee
Contact: Liza Veras, 617.442.9670

<u>Park Site</u>

PLACE: Dudley Town Common South @ Blue Hill Ave. and Dudley St., Roxbury
DATES: Thursdays from June 17 through late October
TIME: 4 pm to 7 pm market (Set-up at 3 pm and Clean-up until 7:30 pm)
VENDORS: 3 Area Farmers; 2 Neighborhood Growers; 2-5 Local Chefs

<u>Products Sold</u>

Home grown fresh produce, flowers, herbs, honey, eggs, and baked goods.

<u>Activities Planned</u>

Throughout the market season, we will highlight healthy food, people, and environment. This may include a health fair, blood pressure screenings, physical fitness testing, line-dancing, nutrition talks, sale of organic garden amendments, cooking demonstrations, urban ecology, gardening, art activities for children, and recycling demonstrations.

We hope to offer presentations by local youth on organic methods of growing food, and invite other youth groups doing theater, dance, music, art, singing, and writing.

<u>Site Use and Visual Impact</u>

Displays

Farmers and Growers will set-up sales displays on the tiers of the South Town Commons. Most farmers erect a free-standing tent and tables near their produce truck or van. The Food Project will provide Neighborhood Gardeners space within our sales area as adjunct vendors.

Water
We recommend that vendors provide their own water for drinking, cleaning produce, etc.

Electricity
We hope to get access to electricity on this site for cooking, music, dance, etc.

Trash
We and other vendors will clean our sales areas following the Market and haul garbage out with us.

Parking: Customers have access to the city lot across the street on Blue Hill or around the corner on Dudley St. Aeromatic Janitorial Supply next door has also approved use of their lot after business hours.

Promotion and Advertising

Department of Food and Agriculture
As an established market with the DFA, we are included in their yearly brochure and marketing campaign.

Farmers Market Coupons
Nuestra Comunidad distributes Farmers' Markets coupons to residents in the Market area. WIC Programs and Neighborhood Health Centers receive mailings to inform their clients/patients about our offerings and location.

Local Businesses
Express support for the Market and will display our poster in their windows. We hope to have several participate or sponsor events.

Local Media
We place notices in community calendars of local papers and radio and advertise in several languages in our flyers to let people know about the work of the Food Project and the existence of this Farmers' Market. There will be a big push in 1999 to invite local residents to the market on a weekly basis via distribution of flyers.

Attachment 7

Dudley Town Common Farmers' Market

Na Merkadu di Kanpunes

El Mercado de Cultivadores

Fresh, Delicious and Affordable!

Tuesdays and Thursdays, 4-7 PM
June - October 2001
Corner of Blue Hill Ave and Dudley St., Roxbury

WIC and Farmers' Market Coupons!

THE FOOD PROJECT

The Food Project Office
(617) 442-1322

¡Que Fresco!

Attachment 8

Compre en el Mercado en Dudley Town

En la esquina de la Avenida Blue Hill y Calle Dudley, Roxbury

Los Martes y Jueves de 4-7 de la tarde
Junio - Octubre, 2001

Seguros ~ cultivados sin pesticidos quimicos
Localmente ~ crecidos por los jovenes de la region
Frescos ~ cosechados y vendidos el mismo dia

Aceptamos WIC !

THE FOOD PROJECT

Por mas información:
The Food Project Office
(617) 442-1322

Attachment 9

Enjoy the Harvest

Dudley Town Common

Farmers' Market

Fresh Convenient Affordable

June - October, 2002
Tuesdays & Thursdays, 4 PM - 7 PM
Corner of Dudley and Blue Hill Ave., Roxbury

$ 5

This flyer allows for the purchase of $5 worth of vegetables, bread, and fruit at our market!

$ 5

WIC Coupons Gladly Accepted!

THE FOOD PROJECT

For More Information:
Call The Food Project Office
(617) 442-1322

May 18, 2000

Calendar Director
Bay State Banner
Boston, MA
Fax 542-7119

Dear Director,

Thank you for listing this announcement in the calendar section of the Bay State Banner. The farm season has just begun and it is an exciting time to be offering fresh produce to the residents here in Roxbury and Dorchester. This year we will be running two farmers' markets at the Dudley Town Common in Roxbury on Tuesdays and Thursdays from 4 to 7 PM. If you could run this announcement for two weeks, that would be great!

Sara Coblyn
The Food Project
781-259-8621

**ANNOUNCEMENT FOR CALENDAR SECTION
OF THE BAY STATE BANNER**

Come buy locally grown and fresh produce at The Dudley Town Common Farmers' Market in Roxbury. Grown by youth from your community on farms in your community!

The Farmers' at Dudley Town Common
The corner of Blue Hill and Dudley in Roxbury
Tuesdays and Thursdays, 4:00 – 7:00 PM
Begins June 1, 2000 through October

For more information,
Contact The Food Project, Sara Coblyn, 781-259-8621 Ext.36

Attachment 11

May 31, 2000

Greater Boston Radio Group
PSA for The Food Project
Fax 617-822-6759

The Dudley Town Common farmers' market operated by The Food Project is an ongoing event that begins on Thursday, June 1, 2000 and runs through October. If you could run this PSA as soon as possible, that would be great. If possible, it would be helpful to run the PSA again in a couple of weeks as the season progresses.

Thanks,
Sara Coblyn
Farmers' Market Manager 781-259-8621

PSA

The Food Project, a local non- profit youth development and sustainable agriculture organization is now beginning its fifth season at the Dudley Town Common Farmers' Market in Roxbury. The market runs on Tuesdays and Thursdays from 4 to 7 PM and will have a variety of organic and locally grown fruits and vegetables as well as transplants, honey, eggs and baked goods. Located at the corner of Blue Hill and Dudley St. in Roxbury across from St. Patrick's Church, on bus lines 15 and 45. Begins June 1 and runs through October. Come to the market and support our local farmers, support our young people that run the market and put your money back into the local community!

March 29, 1999

Dear Director,

Happy Spring from all of the staff at The Food Project! As we plow our fields to prepare them for spring planting, I wanted to touch base with you regarding the approaching farmers' market season. For seven years, The Food Project has run a successful farmers' market at Dudley Town Common in Roxbury, providing fresh, organic produce to residents of the neighborhood. Farmers' Market and WIC coupons from your distribution centers account for over 60% of our sales each season and we want to continue to provide residents with information about our markets at Dudley Town Common. This year we will be running two markets at the Common, on Tuesday and Thursday.

Last year, we designed a flyer for WIC Directors to include in the summer coupon packets for coupon recipients, and are hoping to do so again this year. We would like to include you in this distribution, if you are interested. If this is not possible, I would love to hear of other suggestions that might allow people better access to information about the market. I will follow up with you next week regarding the possibility of this collaboration. Thank you for your time. Enjoy the beautiful weather!

Warm regards,

Colleen O'Brien
Market Staff
(781) 259-8621

Attachment 13

June 1, 1999

Dear WIC Director,

Thank you for distributing these flyers to recipients of WIC and Farmers' Market coupons for the summer season! We are excited to kick-off our fourth season at the Dudley Town Common, Roxbury. We hope that residents will visit the markets each week to share in the harvest. Remember that we will be doing two markets at the common, one on Tuesday and one on Thursday.

We will have plenty of fruits, vegetables, meats, and breads at the market this year as well as chefs, nutritionists, and farmers on staff to share their expertise on the value of locally grown and fresh produce. I have enclosed flyers as well as an original, in case you need to run more copies (Please contact me, if you would like us to send more flyers to you. We would be glad to do this).

Good luck with your work in the coming months. Please call Colleen O'Brien, if you have questions or concerns about the flyers. Thanks again!

Regards,

Colleen O'Brien
Dudley Town Common Market Manager
617-442-1322 or Voicemail 781-259-8621

Attachment 14

Dudley Town Common

Farmers' Market

Fresh

Convenient

Affordable

FIRST MARKET : JUNE 7th

June - October, 2001
Tuesdays & Thursdays, 4 PM - 7 PM
Corner of Dudley and Blue Hill Ave., Roxbury

Everyone Welcome!

WIC Coupons Gladly Accepted!

THE FOOD PROJECT

For More Information:
Call The Food Project Office
(617) 442-1322

Attachment 15-1

March 9, 2000

Dear Potential Vendor:

Happy New Year! I hope it has been a good transition for you and you are looking forward to the 2000 growing season with anticipation.

The Dudley Town Commons Farmers' Market is a vibrant market in Roxbury that has been run by The Food Project on Thursdays from 4-7pm since 1996. Each year our continued presence and outreach efforts have grown this market consistently. In 2000, our presence will be further increased with the addition of another market at the same site on Tuesdays also from 4-7pm. I would like to invite you to participate in one or both of these markets for the 2000 season. This will be an exciting year with the addition of the second market as well as other developments including the sale of fresh products and salsa at our market and an increase in the number of festivals and community wide events at the market. Hopefully, you will be able to attend and add to this diversity.

The Food Project is a youth development organization that brings together young people of diverse backgrounds to grow and distribute food in greater Boston, practicing care for land and community. Through farming, we teach youth critical lessons about diversity, service, self-discipline, small business skills, and the environment. Our youth learn to farm on our 21-acre farm in Lincoln and 2 acre urban agriculture sites in Roxbury. They then follow the seed to the market as our youth participants sell the food they grew at the Dudley Town Commons Farmers' Market. Through structured training sessions, our young people learn to be professionals as they sell at this market.

The Dudley Town Common site is at the center of the Dudley neighborhood. The location is highly visible and busy (as it is a bus stop for folks on public transportation and a throughway for folks in cars) and the site is large enough for several booths. Public parking is also available.

The Food Project has a strong commitment to the Roxbury community and is involved in a grassroots effort began several years ago to transform this neighborhood. And it is happening. The community is changing; there is an increase in jobs, housing and community programs and the movement of young professionals into this area. Development is happening in Dudley Square, increasing the amount of businesses there. The Dudley Town Common Farmers' Market is part of this effort and has grown consistently over the last four years with locally grown produce sold by The Food Project, two area farmers, and one neighborhood gardener. We have regular customers from the neighborhood and newcomers each week. It's a friendly market, with our youth participants engaging in warm conversations with residents about the produce, the weather, and the community. Please think about joining this venture as it continues to grow with this unique community of people.

We need to know ahead of time who will be participating in the market so that we can best organize space and parking availability at the Town Commons. Therefore, please review the enclosed information and submit an application, if you are interested.

I hope you consider trying out this exciting marketplace. There is truly no other experience like that of selling at a vibrant market with people of diverse backgrounds, with young people at the center of it all, learning valuable job skills as well as the importance of supporting locally produced food.

I would be happy to talk further about our market and its location with you and share our sales records if that would help with your decision process. Also, on the application are a couple of options for your participation including a festival day in late summer. We very much hope to hear from you soon and wish you the best in your enterprise. Please don't hesitate to call me.

Warm regards,

Sara Coblyn
Market Manager
617-442-1322
(781) 259-8621

Encl.: Information for Vendors
Vendor Application

Attachment 16-1

2000 Vendor Application & Guidelines

The Dudley Town Common Farmers' Market
Roxbury, MA

Application Due Date: April 20, 2000

Vendor Information

Name of Vendor:
Vendor Contact Person:
Name of Business/Farm/Garden (if applicable):
Mailing Address:

Work Phone: Home Phone:

Emergency Contact Name and Phone #:

Site Plans and Needs

Size of Market Stand/Area needed to sell:

Infrastructure Needs (i.e.. electricity, water):

Market Crops for 2000 Season	Expected Harvest Date

Intentions for Promotion, if any:

Information for Market Management in 2000

We want to make the market an exciting and profitable experience for every vendor who comes to Dudley Town Common, Roxbury. Please answer all the questions below. Applications are due April 20, 2000.

Can you attend for the full market season? Can you attend the first market, June 1?

Specify whether you will be attending the Tuesday or Thursday market or both.

Can we use your name and/or business name when advertising the market?

Are you selling at any other farmers' markets in the City?

What method do you use for pricing your produce/food items?

Do you have any obstacles for selling at the market (i.e.. do not have a scale, table, cash box etc.)?

If you cannot be a continual vendor would you be interested in:

- Attending our "Celebration of local food and farming" event on August 17?
- Selling a certain amount of your product to The Food Project for resale?
  If yes, would you be able to deliver to our Lincoln farm?

Thank you!
Sara Coblyn
Market Manager

Attachment 17

May 20, 2002

John Smith
Dorchester Bay CDC
594 Columbia Road, Suite 302
Dorchester, MA 02125

Dear John:

Greetings from The Food Project!

This letter is to announce the coming opening of our farmers' market in the first week of June. The market has been run for the past six seasons and is located on the Dudley Town Common (the corner of Blue Hill Avenue and Dudley Street in Roxbury). It runs on Tuesday and Thursday afternoons from 4:00pm-7:00pm, June through late October. We would love to talk with you about ways to generate more local interest and participation in the market.

One way to build awareness of local food would be to have our youth make a presentation about the market to your residents sometime in June, whether at a meeting or another event. I look forward to hearing from you when such a gathering will be taking place. In the meantime, I am enclosing some flyers that you can post to inform residents about the market. Thanks very much.

Best Regards,

Kristin Brennan
Market Manager
617-442-1322 x12

Attachment 18-1

Elder Pickup Schedule and Route

Tuesdays between 4:00 and 4:30:

1. Council Towers

Directions to Council Towers, 2875 Washington St., Roxbury
Contact: Frances Sligh, 442-3081

From Smith House on Ruggles, cross Shawmut and follow Ruggles to the intersection with Washington Street. Take a right onto Washington St.. Go straight through the intersection of Dudley and Washington, and follow street up a gradual hill. After passing through the intersection of M.L. King and Washington (there is a light there), you will see a light ahead (the intersection of Washington and Marcella). Go straight. Council Towers is on the right just after this intersection.

To return to market: Take a left back onto Washington and follow to Dudley. Take a right on Dudley.

2. Garrison Apartments

Directions to Garrison Apartments, 60 Brookledge St., Dorchester
Contact: Minnie Simmons, 442-3059

From Council Towers, continue on Washington to the intersection of Washington and Columbus. Take a left (a very busy intersection, be careful) and continue straight. At some point Columbus turns into Seaver, so don't be startled that you've taken a wrong turn. Just continue on. You will eventually see Elm Hill Ave. on your left. Take a left onto Elm Hill Ave. and an immediate left onto Brookledge. 60 Brookledge is on the right, brick building, and you can pull into the parking lot on the right.

To return to market: Go to the end of the straight of Brookledge. Take a right onto Humboldt. Follow to M.L. King. Take a right onto M.L. King and then a left onto Warren. Follow Warren to Moreland. Take a right onto Moreland. At the end, take a left onto Blue Hill. Market is on the right.

Thursdays between 4:00 and 4:30

1. Robert Fortes House

Directions to Robert Fortes House, 541 Shawmut Ave., Roxbury
Contact: Rosetta Peele, #308, 262-2114 or 267-8339

From the market site, take Dudley St. back toward Dudley Station. Just before the Bus Station, take a right onto Warren Street. It quickly turns into Washington St. moments later. Follow this to Melnea Cass. Take a left on Melnea Cass and then a right onto Tremont. At the intersection of Mass. Ave. and Tremont, take a right onto Mass. Ave. Take a right onto Shawmut Ave. Moments later you will see a light at the intersection of Shawmut and Northhampton. Robert Fortes House is located just before that light on the right.

Smith House

Directions to Smith House, 122 Dewitt Dr., Roxbury
Contact: Pauline Sheridan, 427-9242

From Robert Fortes House, follow Shawmut (one way) to Melnea Cass. Cross through the Melne Cass/Shawmut intersection to the next set of lights. Go straight through this next intersection (Ruggles and Shawmut) and take your first right onto Estabrook Rd. Circle around the block to the end and make a right onto Dewitt Dr and then a right again onto Ruggles. The front door of the Smith House is on the right (big brick building. Park beside the handicap ramp.

To return to market: Take Ruggles to Washington. Take a right on Washington and a left on Dudley.

Hey Kids! Tell your parents!

Dudley Town Common

Farmers' Market

WIC Coupons Accepted

Delicious veggies, fruits, and bread!

Fresh, local, and low-prices

June - October, 2001
Tuesdays & Thursdays, 4 PM - 7 PM
Corner of Dudley and Blue Hill Ave., Roxbury

THE FOOD PROJECT

The Food Project, 555 Dudley St.,
Dorchester, MA; (617) 442-1322

Attachment 20

Market Visit Question Sheet

Read the following questions as you ride in the van to the farmers' market. Add any other questions you might be curious about. Be prepared to observe people, set-up, and products at the market. Respectfully ask questions of vendors, customers, and the market manager. At the end, you will report back to your fellow crewmembers about what you observed and learned. From these observations, your crew may suggest certain changes to the staff for the market.

Questions about Display and Product:

- At this market, how are the vegetables displayed?
- Is this market set up any different than ours? If so, in what ways?
- In which ways does this market differ from ours? Do they sell items that we do not?
- Compare the way that we display our fruits and vegetables to the way that this market is displayed? Is the goal of the market to have the items look pretty or to just have the items on the table so that they can be seen by customers?

Questions about Prices and Customers:

- Investigate the prices at the market. Are they higher or lower than ours? If they are higher, why so you think so? Think about those who live in the area and the economic situation of people in the area.
- Did you see any WIC or Farmers' Market coupons being used? If not, why?
- How would you describe some of the people that you saw purchasing products at the market? (Consider race, culture, age, economic class)

Questions for Farmers' and Vendors:

- Are all the vegetables at this market organic? If they are not, why do you think that pesticides are used?
- Where are all these vegetables grown? (This is a good question to ask the vendors.

Set- Up List and Sequence

There are two set up teams, each led by a leader. The market manager oversees both groups. One group sets up the tents and tables (Assistant Leader plus 4), the other is the advertising group (Leader plus 4).

1. Everyone unloads the van. Get the tents and tables out first.

2. **Tents/Tables group** focuses on unloading the tents and tables. When all the tents are out, that group starts to set up the tents while the other group continues to unload everything. Set up the tents so that the middle of them is in line with the black light pole. (We need to leave room for the tables and space behind the tables to work.)

3. **Advertising group** continues to unload the van and put everything neatly behind the tents.

   -2 people set up banner on clock
   -1 person brings sandwich board in front of clock facing road
   -Once tents are up, hang scales, chalkboard, hang bags and set up bin with rest of bags in center of stand, behind tables
   -Prop other signs against tent poles

4. **Tents/Table group** moves to display. Make it pretty, simple and clean!
   -Set up tables
   -Left table is greens with metal bins with water for greens
   -Make sure all overstock is under the table that the vegetable is on and all veggies are in the shade.

5. **Advertising group**
   -Puts signs on vegetables
   -Gets out receipt pads and pens for everyone

6. **4:00 Market** starts. Make sure in weights are done. If there is anyone handing out flyers, they leave with the crew leader. They come back in 30 minutes and rotate with another pair who goes in a different direction.

7. Halfway through the market when it is slow (5:45) the leader manages dinner shifts. Youth must either go sit in the pizza place or come back and sit in the grassy area behind the common. They must not sit near the young people who are operating the market to prevent distractions.

Attachment 22

The Food Project Farmers' Market Sales Log

Location:

Vegetable/Type	L/Quantity	R/Quantity	L/In	R/In	Total	End Quantity	End Weight	Amount Sold	Price
Basil									
Beans, Bush Fr.									
Beans, Bush Snap									
Beans, Shell									
Beets									
Bok Choy									
Broccoli									
Brussels' Sprouts									
Cabbage, Green									
Cabbage, Red									
Cantaloupe									
Carrot									
Cauliflower									
Celeriac									
Chard, Swiss									
Cilantro									
Collards									
Corn, Sweet									
Corn, Pop									
Cucumber, Pick.									

The Food Project Farmers' Market Sales Log

Location:

Date:

Vegetable/Type	L/Quantity	R/Quantity	L/In	R/In	Total	End Quantity	End Weight	Amount Sold	Price
Cucumber, Slic.									
Daikon									
Dill									
Eggplant									
Fennel									
Garlic									
Kale									
Leeks									
Lettuce, Head									
Lettuce, Loose									
Melon, Water									
Mustard Greens									
Onions									
Parsnip									
Peas, Shell									
Peas, Snap									
Peas, Snow									
Pepper, Bell									
Pepper, Multi									
Pepper, Hot									
Potato									

Attachment 23

The Food Project Farmers' Market Sales Log

Location: | Date:

Vegetable/Type	L/Quantity	R/Quantity	L/In	R/In	Total	End Quantity	End Weight	Amount Sold	Price	Final $
Potato, Sweet										
Radish										
Rutabaga										
Salad Mix										
Scallions										
Spinach										
Squash, Patty										
Squash, Yellow										
Squash, Zucchini										
Squash, Winter										
Strawberry										
Tomato, Cherry										
Tomato, Plum										
Tomato, Slicing										
Turnip										
Tomatillo										
Okra										
Celery										
Soybeans										
Turnip Greens										
Mustard Greens										
Beet Greens										
Parsley										
Watermelon										
Flowers										

Attachment 24

Attachment 25

Location:	The Food Project Farmers' Market Accounting Sheet				Date:	
WIC Coupons	x 2.50=	$				
FM $2.50 Coupons	x 2.50=	$				
FM $1.00 Coupons	x 1=	$		CLOSING BALANCE		$
FP Coupons	x 5=	$		OPENING BALANCE	—	$
Cash Total		$		COSTS/OVERHEAD	—	$
Change Total	+	$		TOTAL NET INCOME	=	
CLOSING BALANCE						
Cashbox Start		$				
OPENING BALANCE						
Costs		$		# of Customers:		
Costs		$		Name of Crew:		
Costs		$		Bookkeepers:		
COSTS/OVERHEAD				Intern Initials:		

Attachment 26

Farmers' Markets
Statistical Summary

Market Number	date	Total # Received	Total # Donated	Total # Sold	Total $ Received	Total $ Donated	Total $ Sold
1	6/7/01	136.5	68	68.5	$157.00	$76.25	$80.75
2	6/12/01	109	45	64	$150.75	$30.75	$120.00
3	6/14/01	139	62	44	$168.50	$58.75	$109.75
4	6/19/01	226	93	133	$308.50	$85.00	$223.50
5	6/21/01	193.25	76	117.5	$191.25	$51.00	$140.25
6	6/26/01	353.5	103	250.5	$366.25	$112.00	$254.25
7	6/28/01	487	250.5	226.5	$606.50	$194.25	$412.25
8	7/3/01	430.5	142.5	288	$424.50	$152.25	$272.25
9	7/5/01	910.5	790	120.25	$615.87	$271.50	$344.37
10	7/10/01	489.5	219	272.5	$718.00	$134.50	$583.50
11	7/12/01	693.25	193	500.25	$699.00	$181.00	$518.00
12	7/17/01	602.75	190	412.75	$640.00	$185.75	$454.25
13	7/19/01	513	33	492	$599.25	$27.00	$572.25
14	7/24/01	797.5	228	579.5	$854.25	$252.00	$602.25
15	7/26/01	703.5	217	486.5	$844.50	$231.50	$613.00
16	7/31/01	995.5	267	728.5	$1,102.75	$270.00	$832.75
17	8/2/01	1632	221.5	1410.5	$1,425.25	$246.25	$1,179.00
18	8/7/01	1194	396	850	$1,348.25	$368.00	$980.25
19	8/9/01	1537.5	187	1350.5	$1,650.50	$189.50	$1,461.00
20	8/14/01	1868.5	909.75	958.75	$1,745.25	$631.00	$1,114.25
21	8/16/01	2836.5	646	2190.5	$1,742.25	$574.00	$1,168.25
22	8/21/01	2845	826	2019	$2,189.75	$687.00	$1,502.75
23	8/23/01	3954.5	1120	2834.5	$2,515.50	$1,126.00	$1,389.50
24	8/28/01	2452.5	512	1940.5	$1,787.75	$515.00	$1,272.75
25	8/30/01	2358	304	2054	$1,507.25	$316.00	$1,191.25
26	9/4/01	1442	453	989	$1,492.75	$455.00	$1,037.75
27	9/6/01	2096	192.5	1903.5	$1,448.25	$204.50	$1,243.75
28	9/11/01	221	221	0	$230.75	$230.75	$0.00
29	9/13/01	1232	256	975.5	$1,051.50	$259.50	$792.00
30	9/18/01	2582	367.5	2214.5	$2,026.00	$347.00	$1,679.00
31	9/20/01	1159	402	757	$1,025.25	$356.50	$668.75
32	9/25/01	1044	239	805	$1,021.25	$256.50	$764.75
33	9/27/01	818.1	250	568.1	$816.50	$273.00	$543.50
34	10/2/01	981.5	255	726.5	$827.50	$252.75	$574.75
35	10/4/01	759.25	306	453.25	$721.75	$314.00	$407.75
36	10/9/01	846.5	227	619.5	$765.50	$237.50	$528.00
37	10/11/01	696	95.2	600.8	$648.00	$102.90	$545.10
38	10/16/01	631.5	219	412.5	$645.13	$229.00	$416.13
39	10/18/01	520	127	393	$511.00	$123.00	$388.00
40	10/23/01	734	294.75	439.25	$651.50	$334.00	$317.50
41	10/25/01	446.5	45	401.5	$404.75	$50.50	$354.25
42	10/30/01	1606.5	300	1306.5	$926.75	$75.00	$851.75
TOTALS		46275	12349	33958	$39,572.75	$11,067.65	$28,505.10

Market Receipts

Attachment 27

Market Number	date	Total Cash (data entry)	% of Mkt Sales In cash	Total Coupons (data entry)	% of Mkt Sales in Coupons	Total Received	Number of Market Customers (data entry)
1	6/7/01	$43.00	95%	$2.50	5%	$45.50	13
2	6/12/01	$95.00	86%	$15.00	14%	$110.00	18
3	6/14/01	$103.00	77%	$30.00	23%	$133.00	21
4	6/19/01	$109.00	78%	$30.00	22%	$139.00	30
5	6/21/01	$112.00	92%	$10.00	8%	$122.00	36
6	6/26/01	$156.00	78%	$45.00	22%	$201.00	55
7	6/28/01	$126.00	54%	$107.50	46%	$233.50	42
8	7/3/01	$179.45	75%	$60.00	25%	$239.45	61
9	7/5/01	$116.50	77%	$35.00	23%	$151.50	32
10	7/10/01	$138.00	69%	$62.50	31%	$200.50	60
11	7/12/01	$187.44	60%	$125.00	40%	$312.44	82
12	7/17/01	$160.00	52%	$145.00	48%	$305.00	72
13	7/19/01	$246.50	48%	$270.00	52%	$516.50	124
14	7/24/01	$258.44	48%	$280.00	52%	$538.44	97
15	7/26/01	$316.00	54%	$265.00	46%	$581.00	131
16	7/31/01	$312.00	48%	$342.50	52%	$654.50	108
17	8/2/01	$360.00	43%	$482.50	57%	$842.50	135
18	8/7/01	$214.00	36%	$375.00	64%	$589.00	111
19	8/9/01	$236.00	46%	$277.50	54%	$513.50	123
20	8/14/01	$357.00	44%	$452.50	56%	$809.50	183
21	8/16/01	$430.00	48%	$462.50	52%	$892.50	172
22	8/21/01	$215.00	40%	$322.00	60%	$537.00	160
23	8/23/01	$212.30	24%	$682.50	76%	$894.80	152
24	8/28/01	$283.50	37%	$477.50	63%	$761.00	143
25	8/30/01	$313.90	35%	$587.50	65%	$901.40	188
26	9/4/01	$144.00	37%	$247.50	63%	$391.50	95
27	9/6/01	$346.00	52%	$322.50	48%	$668.50	168
28	9/11/01	$0.00	0%	$0.00	0%	$0.00	0
29	9/13/01	$323.50	55%	$265.00	45%	$588.50	128
30	9/18/01	$190.45	33%	$395.00	67%	$585.45	124
31	9/20/01	$230.65	38%	$376.50	62%	$607.15	112
32	9/25/01	$43.05	12%	$312.50	88%	$355.55	70
33	9/27/01	$217.20	46%	$260.00	54%	$477.20	101
34	10/2/01	$37.00	14%	$235.00	86%	$272.00	65
35	10/4/01	$87.15	25%	$266.50	75%	$353.65	65
36	10/9/01	$71.00	25%	$217.50	75%	$288.50	49
37	10/11/01	$105.00	41%	$151.00	59%	$256.00	51
38	10/16/01	$101.00	34%	$193.50	66%	$294.50	45
39	10/18/01	$88.00	28%	$230.00	72%	$318.00	58
40	10/23/01	$120.00	38%	$192.50	62%	$312.50	37
41	10/25/01	$93.00	25%	$280.00	75%	$373.00	67
42	10/30/01	$33.00	5%	$602.50	95%	$635.50	127
TOTALS		$7,510.03	48%	$10,492.00	52%	$18,002.03	3,711

Attachment 28

Lincoln Statistics

LINCOLN							
Market Number	date	#'s sent to Market	#'s Sold	# Donated	$'s sent to Market	$'s Sold	$'s Donated
1	6/7/01	0	0.00	0.00	$0.00	$0.00	$0.00
2	6/12/01	0	0.00	0.00	$0.00	$0.00	$0.00
3	6/14/01	0	0.00	0.00	$0.00	$0.00	$0.00
4	6/19/01	0	0.00	0.00	$0.00	$0.00	$0.00
5	6/21/01	0	0.00	0.00	$0.00	$0.00	$0.00
6	6/26/01	66	46.77	19.23	$124.50	$86.43	$38.07
7	6/28/01	241	182.85	58.15	$377.50	$256.59	$120.91
8	7/3/01	119.5	80.22	39.28	$161.00	$103.26	$57.74
9	7/5/01	589	403.99	185.01	$360.00	$201.30	$158.70
10	7/10/01	217	158.48	58.52	$445.50	$362.05	$83.45
11	7/12/01	350	256.60	93.40	$539.50	$399.80	$139.70
12	7/17/01	371	254.36	116.64	$421.50	$299.17	$122.33
13	7/19/01	359	342.90	16.10	$439.50	$419.70	$19.80
14	7/24/01	482.5	334.27	148.23	$396.75	$279.71	$117.04
15	7/26/01	546	366.33	179.67	$661.00	$479.80	$181.20
16	7/31/01	666	487.37	178.63	$762.50	$575.81	$186.69
17	8/2/01	1328	1135.96	192.04	$1,102.50	$912.01	$190.49
18	8/7/01	821	585.84	235.16	$971.50	$706.33	$265.17
19	8/9/01	1218	1069.86	148.14	$1,329.50	$1,176.86	$152.64
20	8/14/01	1431.5	948.08	483.42	$1,317.50	$841.15	$476.35
21	8/16/01	2589	2065.08	523.92	$1,518.75	$1,018.38	$500.37
22	8/21/01	2400	1820.46	579.54	$1,807.00	$1,240.08	$566.92
23	8/23/01	3637	2601.40	1035.60	$2,226.50	$1,229.86	$996.64
24	8/28/01	1978	1564.25	413.75	$1,309.00	$931.91	$377.09
25	8/30/01	2097	1817.76	279.24	$1,277.50	$1,009.67	$267.83
26	9/4/01	1239.5	850.11	389.39	$1,293.50	$899.23	$394.27
27	9/6/01	1958	1766.96	191.04	$1,308.00	$1,123.30	$184.70
28	9/11/01	0	0.00	0.00	$0.00	$0.00	$0.00
29	9/13/01	1052	832.98	219.02	$856.00	$644.75	$211.25
30	9/18/01	2321	2009.53	311.47	$1,762.50	$1,460.63	$301.87
31	9/20/01	988	666.62	321.38	$835.50	$544.98	$290.52
32	9/25/01	655	484.98	170.02	$612.00	$458.29	$153.71
33	9/27/01	588	408.32	179.68	$617.50	$411.04	$206.46
34	10/2/01	673	491.29	181.71	$538.75	$374.20	$164.55
35	10/4/01	525	313.41	211.59	$485.00	$274.00	$211.00
36	10/9/01	486	355.67	130.33	$391.00	$269.69	$121.31
37	10/11/01	487	420.39	66.61	$427.00	$359.19	$67.81
38	10/16/01	336	219.48	116.52	$336.00	$216.73	$119.27
39	10/18/01	276	208.59	67.41	$253.00	$192.10	$60.90
40	10/23/01	533	318.96	214.04	$452.00	$220.28	$231.72
41	10/25/01	261	234.70	26.30	$211.00	$184.67	$26.33
42	10/30/01	1396	1096.00	300.00	$721.00	$662.65	$58.35
TOTALS		35281	27201	8080	$28,648.75	$20,825.61	$7,823.14

Boston Statistics **Attachment 29**

BOSTON							
Market Number	date	#'s sent to Market	#'s Used at Market	# Donated	$'s sent to Market	$'s Used at Market	$'s Donated
1	6/7/01	136.5	68.50	68.00	$157.00	$80.75	$76.25
2	6/12/01	109	64.00	45.00	$150.75	$120.00	$30.75
3	6/14/01	139	77.00	62.00	$168.50	$109.75	$58.75
4	6/19/01	226	133.00	93.00	$308.50	$223.50	$85.00
5	6/21/01	193.25	117.25	76.00	$191.25	$140.25	$51.00
6	6/26/01	287.5	203.73	83.77	$241.75	$167.82	$73.93
7	6/28/01	246	186.65	59.35	$229.00	$155.66	$73.34
8	7/3/01	311	208.78	102.22	$263.50	$168.99	$94.51
9	7/5/01	321.5	220.51	100.99	$255.87	$143.07	$112.80
10	7/10/01	272.5	199.02	73.48	$272.50	$221.45	$51.05
11	7/12/01	343.25	251.65	91.60	$159.50	$118.20	$41.30
12	7/17/01	231.75	158.89	72.86	$218.50	$155.08	$63.42
13	7/19/01	154	147.10	6.90	$159.75	$152.55	$7.20
14	7/24/01	315	218.23	96.77	$457.50	$322.54	$134.96
15	7/26/01	157.5	105.67	51.83	$183.50	$133.20	$50.30
16	7/31/01	329.5	241.13	88.37	$340.25	$256.94	$83.31
17	8/2/01	304	260.04	43.96	$322.75	$266.99	$55.76
18	8/7/01	373	266.16	106.84	$376.75	$273.92	$102.83
19	8/9/01	319.5	280.64	38.86	$321.00	$284.14	$36.86
20	8/14/01	437	289.42	147.58	$427.75	$273.10	$154.65
21	8/16/01	247.5	197.42	50.08	$223.50	$149.87	$73.63
22	8/21/01	445	337.54	107.46	$382.75	$262.67	$120.08
23	8/23/01	317.5	227.10	90.40	$289.00	$159.64	$129.36
24	8/28/01	474.5	375.25	99.25	$478.75	$340.84	$137.91
25	8/30/01	261	226.24	34.76	$229.75	$181.58	$48.17
26	9/4/01	202.5	138.89	63.61	$199.25	$138.52	$60.73
27	9/6/01	138	124.54	13.46	$140.25	$120.45	$19.80
28	9/11/01	221	0.00	221.00	$230.75	$0.00	$230.75
29	9/13/01	180	142.52	37.48	$195.50	$147.25	$48.25
30	9/18/01	261	225.97	35.03	$263.50	$218.37	$45.13
31	9/20/01	171	115.38	55.62	$189.75	$123.77	$65.98
32	9/25/01	389	288.02	100.98	$409.25	$306.46	$102.79
33	9/27/01	230.1	159.78	70.32	$199.00	$132.46	$66.54
34	10/2/01	308.5	225.21	83.29	$288.75	$200.55	$88.20
35	10/4/01	234.25	139.84	94.41	$236.75	$133.75	$103.00
36	10/9/01	360.5	263.83	96.67	$374.50	$258.31	$116.19
37	10/11/01	209	180.41	28.59	$221.00	$185.91	$35.09
38	10/16/01	295.5	193.02	102.48	$309.13	$199.40	$109.73
39	10/18/01	244	184.41	59.59	$258.00	$195.90	$62.10
40	10/23/01	201	120.29	80.71	$199.50	$97.22	$102.28
41	10/25/01	185.5	166.80	18.70	$193.75	$169.58	$24.17
42	10/30/01	210.5	171.19	39.31	$205.75	$189.10	$16.65
TOTALS		10993.6	7901	3093	$10,924.00	$7,679.49	$3,244.51

Attachment 30

Market Donation Sheet

Market Number	date	Total #'s Received	data entry Total #'s Donated	Total #'s Sold	Donation % by #	Total $'s Received	data entry Total $'s Donated	Total $'s Sold	Donation % by $
1	6/7/01	136.5	68	68.5	49.82%	$157.00	$76.25	$80.75	48.57%
2	6/12/01	109	45	0	41.28%	$150.75	$30.75	$120.00	20.40%
3	6/14/01	139	62	77	44.60%	$168.50	$58.75	$109.75	34.87%
4	6/19/01	226	93	133	41.15%	$308.50	$85.00	$223.50	27.55%
5	6/21/01	193.25	76	117.25	39.33%	$191.25	$51.00	$140.25	26.67%
6	6/26/01	353.5	103	250.5	29.14%	$366.25	$112.00	$254.25	30.58%
7	6/28/01	487	117.5	369.5	24.13%	$606.50	$194.25	$412.25	32.03%
8	7/3/01	430.5	141.5	289	32.87%	$424.50	$152.25	$272.25	35.87%
9	7/5/01	910.5	286	624.5	31.41%	$615.87	$271.50	$344.37	44.08%
10	7/10/01	489.5	132	357.5	26.97%	$718.00	$134.50	$583.50	18.73%
11	7/12/01	693.25	185	508.25	26.69%	$699.00	$181.00	$518.00	25.89%
12	7/17/01	602.75	189.5	413.25	31.44%	$640.00	$185.75	$454.25	29.02%
13	7/19/01	513	23	490	4.48%	$599.25	$27.00	$572.25	4.51%
14	7/24/01	797.5	245	552.5	30.72%	$854.25	$252.00	$602.25	29.50%
15	7/26/01	703.5	231.5	472	32.91%	$844.50	$231.50	$613.00	27.41%
16	7/31/01	995.5	267	728.5	26.82%	$1,102.75	$270.00	$832.75	24.48%
17	8/2/01	1632	236	1396	14.46%	$1,425.25	$246.25	$1,179.00	17.28%
18	8/7/01	1194	342	852	28.64%	$1,348.25	$368.00	$980.25	27.29%
19	8/9/01	1537.5	187	1350.5	12.16%	$1,650.50	$189.50	$1,461.00	11.48%
20	8/14/01	1868.5	631	1237.5	33.77%	$1,745.25	$631.00	$1,114.25	36.16%
21	8/16/01	2836.5	574	2262.5	20.24%	$1,742.25	$574.00	$1,168.25	32.95%
22	8/21/01	2845	687	2158	24.15%	$2,189.75	$687.00	$1,502.75	31.37%
23	8/23/01	3954.5	1126	2828.5	28.47%	$2,515.50	$1,126.00	$1,389.50	44.76%
24	8/28/01	2452.5	513	1939.5	20.92%	$1,787.75	$515.00	$1,272.75	28.81%
25	8/30/01	2358	314	2044	13.32%	$1,507.25	$316.00	$1,191.25	20.97%
26	9/4/01	1442	453	989	31.41%	$1,492.75	$455.00	$1,037.75	30.48%
27	9/6/01	2096	204.5	1891.5	9.76%	$1,448.25	$204.50	$1,243.75	14.12%
28	9/11/01	221	221	0	100.00%	$230.75	$230.75	$0.00	100.00%
29	9/13/01	1232	256.5	975.5	20.82%	$1,051.50	$259.50	$792.00	24.68%
30	9/18/01	2582	346.5	2235.5	13.42%	$2,026.00	$347.00	$1,679.00	17.13%
31	9/20/01	1159	377	782	32.53%	$1,025.25	$356.50	$668.75	34.77%
32	9/25/01	1044	271	773	25.96%	$1,021.25	$256.50	$764.75	25.12%
33	9/27/01	818.1	250	568.1	30.56%	$816.50	$273.00	$543.50	33.44%
34	10/2/01	981.5	265	716.5	27.00%	$827.50	$252.75	$574.75	30.54%
35	10/4/01	759.25	306	453.25	40.30%	$721.75	$314.00	$407.75	43.51%
36	10/9/01	846.5	227	619.5	26.82%	$765.50	$237.50	$528.00	31.03%
37	10/11/01	696	95.2	600.8	13.68%	$648.00	$102.90	$545.10	15.88%
38	10/16/01	631.5	219	412.5	34.68%	$645.13	$229.00	$416.13	35.50%
39	10/18/01	520	127	393	24.42%	$511.00	$123.00	$388.00	24.07%
40	10/23/01	734	294.75	439.25	40.16%	$651.50	$334.00	$317.50	51.27%
41	10/25/01	446.5	45	401.5	10.08%	$404.75	$50.50	$354.25	12.48%
42	10/30/01	1606.5	300	1306.5	18.67%	$926.75	$75.00	$851.75	8.09%
TOTALS		46275	11133	35077		$39,572.75	$11,067.65	$28,505.10	

Farm % breakdown by $'s

Attachment 31

Market Number	date	Total $'s Received	data entry $'s from Lincoln	% from Lincoln	data entry $'s from Boston	% from Boston
1	6/7/01	$157.00	$0.00	0.00%	$157.00	100.00%
2	6/12/01	$150.75	$0.00	0.00%	$150.75	100.00%
3	6/14/01	$168.50	$0.00	0.00%	$168.50	100.00%
4	6/19/01	$308.50	$0.00	0.00%	$308.50	100.00%
5	6/21/01	$191.25	$0.00	0.00%	$191.25	100.00%
6	6/26/01	$366.25	$124.50	33.99%	$241.75	66.01%
7	6/28/01	$606.50	$377.50	62.24%	$229.00	37.76%
8	7/3/01	$424.50	$161.00	37.93%	$263.50	62.07%
9	7/5/01	$615.87	$360.00	58.45%	$255.87	41.55%
10	7/10/01	$718.00	$445.50	62.05%	$272.50	37.95%
11	7/12/01	$699.00	$539.50	77.18%	$159.50	22.82%
12	7/17/01	$640.00	$421.50	65.86%	$218.50	34.14%
13	7/19/01	$599.25	$439.50	73.34%	$159.75	26.66%
14	7/24/01	$854.25	$396.75	46.44%	$457.50	53.56%
15	7/26/01	$844.50	$661.00	78.27%	$183.50	21.73%
16	7/31/01	$1,102.75	$762.50	69.15%	$340.25	30.85%
17	8/2/01	$1,425.25	$1,102.50	77.35%	$322.75	22.65%
18	8/7/01	$1,348.25	$971.50	72.06%	$376.75	27.94%
19	8/9/01	$1,650.50	$1,329.50	80.55%	$321.00	19.45%
20	8/14/01	$1,745.25	$1,317.50	75.49%	$427.75	24.51%
21	8/16/01	$1,742.25	$1,518.75	87.17%	$223.50	12.83%
22	8/21/01	$2,189.75	$1,807.00	82.52%	$382.75	17.48%
23	8/23/01	$2,515.50	$2,226.50	88.51%	$289.00	11.49%
24	8/28/01	$1,787.75	$1,309.00	73.22%	$478.75	26.78%
25	8/30/01	$1,507.25	$1,277.50	84.76%	$229.75	15.24%
26	9/4/01	$1,492.75	$1,293.50	86.65%	$199.25	13.35%
27	9/6/01	$1,448.25	$1,308.00	90.32%	$140.25	9.68%
28	9/11/01	$230.75	$0.00	0.00%	$230.75	100.00%
29	9/13/01	$1,051.50	$856.00	81.41%	$195.50	18.59%
30	9/18/01	$2,026.00	$1,762.50	86.99%	$263.50	13.01%
31	9/20/01	$1,025.25	$835.50	81.49%	$189.75	18.51%
32	9/25/01	$1,021.25	$612.00	59.93%	$409.25	40.07%
33	9/27/01	$816.50	$617.50	75.63%	$199.00	24.37%
34	10/2/01	$827.50	$538.75	65.11%	$288.75	34.89%
35	10/4/01	$721.75	$485.00	67.20%	$236.75	32.80%
36	10/9/01	$765.50	$391.00	51.08%	$374.50	48.92%
37	10/11/01	$648.00	$427.00	65.90%	$221.00	34.10%
38	10/16/01	$645.13	$336.00	52.08%	$309.13	47.92%
39	10/18/01	$511.00	$253.00	49.51%	$258.00	50.49%
40	10/23/01	$651.50	$452.00	69.38%	$199.50	30.62%
41	10/25/01	$404.75	$211.00	52.13%	$193.75	47.87%
42	10/30/01	$926.75	$721.00	77.80%	$205.75	22.20%
TOTALS		$39,572.75	$28,648.75	60%	$10,924.00	40%

Attachment 32

Farm % breakdown by #'s

	data entry			data entry		data entry		
Market Number	date	Total lbs Received		lbs from Lincoln	% from Lincoln	lbs from Boston		% from Boston
1	6/7/01	136.5		0	0%	136.5		100%
2	6/12/01	109		0	0%	109		100%
3	6/14/01	139		0	0%	139		100%
4	6/19/01	226		0	0%	226		100%
5	6/21/01	193.25		0	0%	193.25		100%
6	6/26/01	353.5		66	19%	287.5		81%
7	6/28/01	487		241	49%	246		51%
8	7/3/01	430.5		119.5	28%	311		72%
9	7/5/01	910.5		589	65%	321.5		35%
10	7/10/01	489.5	#	217	44%	272.5		56%
11	7/12/01	693.25		350	50%	343.25		50%
12	7/17/01	602.75		371	62%	231.75		38%
13	7/19/01	513		359	70%	154		30%
14	7/24/01	797.5		482.5	61%	315		39%
15	7/26/01	703.5		546	78%	157.5		22%
16	7/31/01	995.5		666	67%	329.5		33%
17	8/2/01	1632		1328	81%	304		19%
18	8/7/01	1194		821	69%	373		31%
19	8/9/01	1537.5		1218	79%	319.5		21%
20	8/14/01	1868.5		1431.5	77%	437		23%
21	8/16/01	2836.5		2589	91%	247.5		9%
22	8/21/01	2845		2400	84%	445		16%
23	8/23/01	3954.5		3637	92%	317.5		8%
24	8/28/01	2452.5		1978	81%	474.5		19%
25	8/30/01	2358		2097	89%	261		11%
26	9/4/01	1442		1239.5	86%	202.5		14%
27	9/6/01	2096		1958	93%	138		7%
28	9/11/01	221		0	0%	221		100%
29	9/13/01	1232		1052	85%	180		15%
30	9/18/01	2582		2321	90%	261		10%
31	9/20/01	1159		988	85%	171		15%
32	9/25/01	1044		655	63%	389		37%
33	9/27/01	818.1		588	72%	230.1		28%
34	10/2/01	981.5		673	69%	308.5		31%
35	10/4/01	759.25		525	69%	234.25		31%
36	10/9/01	846.5		486	57%	360.5		43%
37	10/11/01	696		487	70%	209		30%
38	10/16/01	631.5		336	53%	295.5		47%
39	10/18/01	520		276	53%	244		47%
40	10/23/01	734		533	73%	201		27%
41	10/25/01	446.5		261	58%	185.5		42%
42	10/30/01	1606.5		1396	87%	210.5		13%
TOTALS		46274.6		35281	59%	10993.6		41%

Market Analysis Workshop

I. Analysis
   A. Qualitative vs. Quantitative
   B. Other Vocabulary to Explore: Revenue, Gross, Net, Profit, Income, Expenses, Overhead, Profit Margin, Margin of Error

II. Qualitative Analysis
   1. Personal Feedback about Market
   2. Suggestions
   3. Specific Experiences

III. Quantitative Analysis
   A. Inventory of Vegetables In and Out of the Market
   B. Market 1 and 2 of the week
      1. Count cash
      2. Count Coupons
      3. Count # of customers
      4. Figure in costs of market (baked goods, fruit, other items)
      5. Compare receipt amount to net amount
   C. WIC Coupon – Stamping and Counting; Place coupons in groups of 100 to be reimbursed
   D. Inventory of Materials: Pens, Tape, Chalk, Receipt Pads, Rubber Bands (any materials you deem essential to the market)

IV. Regroup:
   A. Have each group of youth describe what they did.
   B. What were the results? How many pounds of vegetables came into the market? How many pounds were sold? How much money was made? In coupons? In cash? How many customers were there?
   C. What did each activity teach you? How did the two markets that the groups ran compare?